GREAT COMMANDS

GREAT WAR COMMANDS

Historical Perspectives on Canadian Army Leadership 1914-1918

Edited By

Andrew B. Godefroy

CANADIAN DEFENCE ACADEMY PRESS

Canadian Defence Academy Press
PO Box 17000 Stn Forces
Kingston, Ontario K7K 7B4

Produced for the Canadian Defence Academy Press
by 17 Wing Winnipeg Publishing Office.
WPO30553

Inside Front & Back Cover Maps: Jay Winter and Blaine Baggett, *The Great War and the Shaping of the 20th Century* (New York: Penguin Books, 1996).

Library and Archives Canada Cataloguing in Publication

Great War commands : historical perspectives on Canadian Army leadership, 1914-1918 / edited by Andrew B. Godefroy.

"Produced for the Canadian Defence Academy Press by 17 Wing Winnipeg Publishing Office."
Available also on the Internet.
Issued by: Canadian Defence Academy.
Includes bibliographical references and index.
ISBN 978-1-100-15600-2 (bound).
ISBN 978-1-100-15601-9 (pbk.)
Cat. no.: D2-259/1-2010E (bound)
Cat. no.: D2-259/2-2010E (pbk.)

1. Canada. Canadian Army. Canadian Expeditionary Force--History. 2. Canada. Canadian Army. Canadian Expeditionary Force--Biography. 3. Canada. Canadian Army. Canadian Corps--History. 4. Military art and science--Canada. 5. Canada--Armed Forces--History--World War, 1914-1918.
6. World War, 1914-1918--Campaigns--Western Front. 7. Command of troops. I. Godefroy, Andrew B., 1972- II. Canadian Defence Academy III. Canada. Canadian Armed Forces. Wing, 17

D547 C2 G745 2010 940.4'1271 C2010-980144-X

Printed in Canada.

1 3 5 7 9 10 8 6 4 2

ACKNOWLEDGEMENTS

This book is the product of several good discussions with a number of the authors appearing within this study. Collectively, we all agreed that there needed to be a greater appreciation for the men who effectively commanded and led the Canadian Expeditionary Force during the First World War, therefore it made sense to pool our efforts to produce such an analysis. Though my name appears on the cover, this book would not have been possible without their efforts.

Initial support for this project came from the Canadian Army's Directorate of Land Concepts and Designs, and the Canadian Forces Leadership Institute of the Canadian Defence Academy. The authors are also indebted to the following institutions for their generous assistance and access to materials – Library and Archives Canada; the Canadian War Museum; the Provincial Archives of Alberta, Ontario, and British Columbia; the Royal Military College of Canada archives; and the Fort Frontenac Army Library. Many photographs appearing in this book were provided courtesy of Library and Archives Canada.

The editor also wishes to thank Colonel Bernd Horn for his publication support, as well as Professor Terry Copp and Michael Bechthold of the Laurier Centre for Military Strategic and Disarmament Studies for their comments and suggestions along the way. Finally, each of the contributors deserves mention for their support to this project, but any errors or omissions in the manuscript are the responsibility of the editor alone.

TABLE OF CONTENTS

Editor's Note

Although the modern use of hyphenation in, and forms of abbreviation of, Canadian Forces ranks have not always been the stylistic norm, in the interests of consistency and universal comprehension, the modern style will used within this book, except where precluded in the case of historical quotes and sources.

FOREWORD

The Canadian Expeditionary Force and its central formation, the Canadian Corps, was an unusual formation for its time. Born out of necessity and shaped by the industrialized demands of the First World War, it bore very little resemblance to any previous war organization ever mobilized by this country. Still, despite its rather unique origin, Canada's Great War army quickly came to grips with the terrible demands of modern industrialized warfare and played a leading role in the Entente's victory on the Western Front. This, despite the fact that its ranks were filled with citizen-soldiers, few of which had substantial previous military training or combat experience, has ensured that the CEF is still considered today as one of the most successful and combat effective formations ever fielded by Canada.

Yet, despite its famous attraction to both scholars and practitioners as a subject of study, much about the CEF and its field formations remains unexplored. For example, we still know very little about those individuals who commanded this incredible formation through some of the worst and most deadly conflicts in the history of modern warfare. Though its most recognizable leader, Sir Arthur Currie, has garnered some attention from biographers, many of these studies are extremely dated and do not consider his actions as a leader in great depth. As for those immediately under his command, few details are known about these senior officers. Even less is known about the "supporting actors", yet it was often the leadership and innovation of these specialists that defeated the tremendous obstacles that trench warfare imposed.

Thus, it is with great pleasure that I introduce to you Andrew Godefroy's *Great War Commands: Historical Perspectives on Canadian Army Leadership, 1914-1918*. Dr. Godefroy has researched and written extensively on the subject of the First World War for over a decade and is an acknowledged subject-matter expert. He has examined the origin, evolution, institution, and combat effectiveness of Canada's Great War army at all levels, and this study represents a tremendous effort on his part to bring together some of the country's leading military historians of the Great War to examine a critical topic of interest in a single volume. *Great War*

FOREWORD

Commands successfully fills a critical gap in our knowledge of the senior leadership of the Canadian Corps through a detailed study of several of its most important figures.

The study of historical leadership and command remains an essential requirement for the successful education and professional development of our armed forces today. This is not because the study of history will necessarily prove anything or validate modern methods, but rather because history is revealing, and its study often shows us what may happen, what is possible, and most important, what questions we may need to ask of ourselves when faced with similar situations today. As in the Great War, at times one can only imagine what challenges the leader may face on the battlefield. This study and others like it will assist greatly in reducing that uncertainty, and hopefully, perhaps even mitigate a small portion of the risk.

Major-General J.P.Y.D. Gosselin
Commander
Canadian Defence Academy

INTRODUCTION

ANDREW B. GODEFROY

From popular literature to reprinted memoirs and new media, over the last decade military historians have taken a renewed interest in Canada's role in the First World War. In particular, their attention has focused greatly on the Canadian Expeditionary Force (CEF) and its decisively lethal Canadian Corps, an ably-led and well-supported combat formation that was often unmatched for success on the western front. With the strength of four full infantry divisions and support arms from 1916 onwards, the Canadian Corps became a hammer for the British Army, soon earning the title "shock troops" and often later referred to as the "tip of the spear" in the drive towards final victory on the western front. By the end of the war, over a half million men and women served in the CEF and the Canadian Corps. Sadly, 64,944 of them never returned home.

Studies of military organizations cannot be considered complete without some consideration for those who lead, shape, and guide them through both war and peace. Yet, despite the renewed attention on the Canadian Corps itself, the study of those who commanded this juggernaut at the highest levels remains much less well defined than the mass of men who filed its ranks. This is somewhat odd given that there exist many detailed political, social, operational, and tactical studies on the CEF, and begs one to ask how historians have assessed the movements and actions of the body of the Canadian Corps without a developed understanding of what was going on in the mind of this titan as it did so.

It is often forgotten, for example, that no less than 126 generals led the Canadian Army during the First World War, with approximately 78 of these senior officers commanding overseas and on the western front. Such numbers were not a reflection of wasteful redundancy, rather, it was the critical mass of leadership required for the formation and fielding of Canada's first large scale national army. Yet of these men only one, Arthur Currie, has merited more than one examination of his professional life. The rest have received much less treatment, while most, including several examined in this book, have yet to receive any historical attention at all.

One cannot help but wonder why this is so when their counterparts have faced much greater scrutiny since the end of the war. British generals were openly and publicly criticized as early as the 1920s, and certainly after 1936 when British Prime Minister David Lloyd George described generals venomously in his *War Memoirs* as a group "regarding thinking as a form of mutiny." Another damning assessment arrived in 1961 when Alan Clark published his infamously titled book *The Donkeys* from which the now common term "lions led by donkeys" is derived when describing the roles of commanders and subordinates in war. The 1960s play and film *Oh What a Lovely War* and its 1989 next generation counterpart *Blackadder Goes Forth* solidified for my own generation that traditional image of red-tabbed idiot commanders and their lackey staff officers sending much better men to a senseless death. One would think that Canadian Great War generals, in turn, would have received at least similar negative popular attention from our own public, not silence.

Perhaps this silence resulted partially from the fact that in the years immediately following the war, the story of the Canadian Corps and the CEF writ large remained untold. The proposed multi-volume official history of the Canadian Army in the Great War languished for years and in the end produced only a single volume covering the first year of battles, a time when Canada fielded only a single infantry division and many of the generals had yet to become more directly involved in operations. Though the gap in literature was eventually filled by others tired of waiting for the historical staff to complete their work, many of these publications were focused at the regimental and battalion level, where the most senior officer mentioned or examined was the lieutenant-colonel commanding.

Individual attempts by general officers to publish memoirs after the war also did not succeed in any great number. Those officers who did not vilify the war in their writings were often criticized by potential publishers for having comforts and privileges that were denied to their men. The fact that those privileges included a much higher chance of being killed in action if identified by enemy snipers or machine gunners as the leader, or that no amount of money could buy off the trajectory of an enemy artillery shell, was conveniently ignored. Once in the line, all men regardless of rank faced equal dangers.

Nor was there much sympathy for the burden command placed on these men's shoulders. Senior officers, both British and Canadian, are typically described as callous and aloof, as cold as their stares in the age-old photos that captured them. Yet is it simply ludicrous to suggest that these men did not feel the pain and guilt of command, which at times must have been immeasurable, and arguably remains little understood by those of us who are nothing more than vicarious witnesses to operational and tactical command in total war. All senior officers knew that even the most successfully executed plans would kill some of their men. This was a terrible price to acknowledge and still carry on. In many ways it may have also assured that the generals sought no attempt to defend themselves or their actions after the war.

Though improved studies of the formation and functionality of the CEF have recently appeared, most are focused on the 1st Canadian Division where there are the richest contemporary sources and the greatest depth. In contrast there is no published account examining the command or operations of the 2nd, 3rd, and 4th Canadian Divisions, only one unpublished brigade history, and no post-1970s modern studies of other major formations or branches. Instead, students and scholars of the period continue to pursue either macro social assessments or micro battlefield analyses, often without the greater context of the Canadian Corps or the British Army in which it served. Added to this dilemma is a generally poor appreciation of the enemy the Canadian Corps fought against. Few Canadian histories mention the Imperial German Army in any detail at all.

Such a state of affairs, combined with a general trend in Canadian history towards avoiding military biography, has ensured that the senior commanders in the CEF and the Canadian Corps remain something of a mystery. As previously mentioned, only Arthur Currie has entertained more than one study of his military career, but the last of these appeared in 1987. None of his immediate divisional level subordinates – Major-Generals Macdonnell, Burstall, Loomis, or Watson – have received even a single dedicated study of their professional careers. Other senior officers still would hardly be recognized, even by Canadian military historians.

For example, very little is known about Brigadier-General Charles Johnstone Armstrong, who served as Chief Engineer of the Canadian Corps or his successor, Brigadier-General William Bethune Lindsay. The same could be said of Brigadier-General Sir Edward W.B. Morrison, the General Officer Commanding Canadian Corps Artillery.

This book seeks to address some of the many existing gaps in the history of the CEF through a study of the most senior leadership commanding the Canadian Corps. It is by no means exhaustive – a detailed study of the Canadian generals would require, if not merit, a monograph of its own. Still, the chapters contributed here greatly broaden our current understanding of how the CEF was led and why it performed as it did both at home and on the battlefield. For the purposes of this study, the editor was fortunate to bring together some of the country's leading scholars on the CEF to examine the nature of its leadership, command, and control. The experience of the contributors range widely from academic to professional soldier, providing a depth of critical examination and understanding to the subject that it has never previously enjoyed.

The first two chapters are devoted to examining two key figures in Canada's Great War army. In the opening chapter, the editor dissects Major-General Arthur Currie's approach to the art of war during the Battle of Mount Sorrel in June 1916. A difficult battle that Currie ultimately won, it was perhaps his greatest test in the year before he took command of the Canadian Corps. Next, David Campbell examines leadership and the fog of war through Major-General Richard Turner's command at the Battle of St. Eloi in early 1916. Unlike Currie, Turner enjoyed no victory at the terrible "battle of the craters", yet Campbell's detailed analysis reveals that pervious criticisms of Turner's performance have misconstrued the challenges he faced.

Following the chapters on Currie and Turner is an analysis of three other senior officers in the corps. Ian McCulloch has provided a detailed view of Major-General Sir Archibald Macdonell's command at both the brigade and division level, focusing on his personal character as well as his influence on his subordinates. In some contrast, David O'Keefe has adopted a grim narrative for Brigadier-General Frederick Loomis in the following chapter, employing the general's only recently rediscovered wartime diary to reveal the very personal burdens faced in carrying the weight of the decisions of command. Finally, Patrick Brennan completes this trio of investigations at the divisional level with his examination of Major-General David Watson, and offers a new interpretation of the strengths and weaknesses of this venerable divisional officer.

Leadership did not only reside at the highest levels of authority. Supporting the senior command were a myriad of intelligent, caring, and resourceful warriors who formed the backbone of innovation necessary to make the Canadian Corps combat effective. Paul Dickson, a leading scholar on General Andrew McNaughton, examines the role of this future general in shaping early Canadian and British counter-battery staff offices. Next, Tim Winegard singles out two battalion commanders who strove to transcend the barriers of Canadian warrior culture with his study of Lieutenant-Colonels Glen Campbell and Andrew Thompson. As the commanding officers of the only two native Amerindian battalions, Campbell and Thompson struggled to have the abilities of the First Nations recognized. Finally, Tod Strickland dissects the professional development of a better-known battalion commander under the spotlight in his innovative study of Lieutenant-Colonel Agar Adamson. A venerable character that left behind detailed personal memoirs, Adamson remains a popular subject in army leadership analysis.

The modern reader may find many similarities between the Great War commands of Canadian soldiers and those who command on the battlefield today. Most important, however, is that the study of leadership and command remains a critical activity for today's military, especially at the formation level and higher. This book contributes to that goal through the examination of the leadership of one of Canada's most historically famous and combat effective formations.

CHAPTER 1

The Advent of the Set-Piece Attack

Major-General Arthur Currie and the Battle of Mount Sorrel
2-13 June 1916

ANDREW B. GODEFROY

As the head of a rather modest pantheon of notable Canadian generals, Sir Arthur William Currie remains one of the most easily recognizable commanders in Canadian military history today. He is also the only senior officer in the Canadian Corps to have merited more than one complete biography of his life, whereas his counterparts as well as those he led in the Great War have for the most part fallen into the shadows.[1] Still, the venerable corps commander himself has received only scant attention from military historians and academics since the publication of Daniel G. Dancock's popular 1985 biography – a less than flattering legacy given his central role in Canada's Great War battles.[2]

Currie's operational performance at various command levels prior to his appointment as General Officer Commanding (GOC) the Canadian Corps in June 1917 are noticeably absent from the existing examinations of his life. This period of Currie's professional development is instead dominated by intrigue, such as the issue of his pre-war financial affairs, or the political debate between official historians Fortescue Duguid and Sir James Edmonds over Currie's personal conduct at the Second Battle of Ypres in April 1915. While these issues certainly lend depth to the complex character of this man, they do less to explain how Currie evolved into arguably the best Corps Commander in the British Expeditionary Force (BEF).[3] Therefore, in order to fully understand Sir Arthur Currie's ascension to Corps Command, it is necessary to evaluate his operational performance at critical engagements prior to this pinnacle appointment.

Arthur Currie began his wartime military career from a place of privilege. As one of only three brigade commanders in the first Canadian contingent overseas in 1914, Currie had an early opportunity to show his tactical acumen and penchant for leadership. At the same time it was also a precarious platform from which he could easily fall, either as a result of his own failings or the by the actions of those he commanded. Despite some initial criticism concerning his abilities, however, Currie executed his duties as brigade commander well enough to be promoted to divisional command in September 1915. From that date until June 1917, Major-General Currie led the 1st Canadian Division in several engagements of varying sizes that severely tested his leadership and command skills. It was during this period that Currie fought some of his toughest battles, and proved that he had both the tactical skill as well as the military-political savvy needed to command at the Corps level the following year.

The Battle of Mount Sorrel in June 1916 constituted Arthur Currie's first real test as a divisional commander, but it has yet to be recognized as a catalyst in his career. None of the three published biographies of his life paid more than passing attention to his command performance during this grueling twelve-day battle. Historians Hugh Urquhart and Daniel G. Dancocks both devoted only three pages to the event while historian A.M.J. Hyatt afforded only a page and a half on Currie's command at this engagement. All of his biographers appear to have concluded, perhaps erroneously, that Currie's divisional command experience had little influence on his subsequent actions as a Corps Commander.

Yet it was at this engagement that the future Corps Commander first executed what became known later as a "set-piece attack" – a deliberately planned offensive tactic consisting of a deliberate closely timed infantry assault under the cover of an intense rolling barrage against a limited objective. Given the limitations of communications and navigation technology of the period, such attacks could be controlled up to about the Corps level, and when properly executed, usually achieved the desired result. This approach, once described by historian Ian Brown as "not glamourous but effective" became such an effective approach in the somewhat unique environment that was the western front that it was soon adopted as the staple of both the Canadian Corps as well as other Entente forces for the remainder of the war.[4]

The Battle of Mount Sorrel is also important to the study of Currie's development as a commander because of the many tests he faced during this engagement. Currie had to start the battle on the defensive and fight to regain the initiative from a dangerous foe. Instead of the typical Prussian *landwehr* or *ersatz* regiments, Currie's division squared off against the *XIII. (Königlich Württembergisches) Armeekorps*, the Imperial German Army's formidable Thirteenth (Royal Württemberg) Army Corps. The initial situation also forced his division into executing a hasty attack with composite formations, and without time to conduct proper battle procedure, this counter-attack achieved little. Yet, when given adequate time to reduce enemy combat effectiveness and prepare for battle, Currie's formations achieved immediate success against his adversary. Most important, perhaps, is that all of these challenges led Currie to not only understand the type of battle he was fighting, but also how to win it against one of the toughest opponents the Imperial German Army could throw at him.

In the early days of fighting on the western front, British generals favouring manoeuvrist tactics that would give the cavalry a central role in defeating the Imperial German Army were highly critical of any infantry-led "bite-and-hold" approaches. Yet given the ability of modern artillery and machine guns to decisively halt any traditional cavalry charges, sweeping horse-led breakthroughs on the western front were unlikely. Instead, the rapid seizure and control of key terrain and vital ground by well-armed and supported infantry could force the enemy back, and when executed in rapid succession, such attacks could prove decisive. At the Battle of Mount Sorrel in June 1916, Arthur Currie applied for the first time this deliberate set-piece attack method, and opened a new door towards victory.

DEFENDING THE YPRES SALIENT

The Ypres salient was considered one of the most important yet deadly sectors of the Entente line on the western front, and defending it against constant German attack proved to be a daily nightmare for the soldiers burdened with the task. The area was scarred from repeated battles and it was the scene of some of the fiercest fighting between British, Canadian, and German forces during the summer of 1915. Most of the structures in the area around Mount Sorrel were long ago reduced to smashed and burnt out rubble, while most of the green fields and woods had been churned into brown splinters and muck. As one historian later commented, "despite being midsummer – June 1916 – the landscape had a wintry feel with only the distant trees being in full leaf, the others being shell-swept and struggling to sprout greenery following the devastating barrages …"[5]

The front lines of Ypres had shifted little since the earlier battles of 1915. In the north of the salient, the British still held ground captured the previous autumn just in front of a small copse of trees named "Y Wood" due to its distinctive shape. Their front line snaked south from here in between the village of Hooge and its chateau. Turning south across the Menin Road, the Canadian front line took over at this point and followed roughly the eastern edges of Sanctuary and Armagh woods, then south again along a series of high features identified as Tor Top (Hill 62), Hill 61, and finally Mount Sorrel.[6] Opposite them, the Germans held tenaciously onto a north-south route known as Green Jacket Ride, as well as a large feature known as Stirling Castle.

In terms of specific unit locations, the Canadian defensive disposition north to south along the Ypres salient on 2 June was as follows. Next to the V (British) Corps boundary to the north was 7th Canadian Infantry Brigade (7 CIB) with companies of the Royal Canadian Regiment (RCR) holding southern area around Hooge and companies of the Princess Patricia's Canadian Light Infantry (PPCLI) forward holding a defensive line through the north end of Sanctuary Wood. The 42nd (Black Watch) and 49th (Edmonton) Battalions belonging to 7 CIB remained in reserve. The 8th Canadian Infantry Brigade was situated to the south of 7 CIB, with the 1st Canadian Mounted Rifles (1 CMR) holding a defensive position through the south end of Sanctuary Wood and along Observatory Ridge to Tor Top. To their right was 4 CMR. This unit, minus one of its companies held in reserve, defended the ridge from Tor Top to the far side of Mount Sorrel. The 3rd Canadian Division boundary touched the 1st Canadian Division boundary just past Mount Sorrel and ran back through Square Wood. To the right of 4 CMR was the 1st Canadian Division's tactical boundary line.

The 1st Canadian Division defended the front line from the southern side of Mount Sorrel southwest towards Hill 60 and the Ypres-Menin railway embankment. On 1 June 1916 Major-General Currie had deployed the 2nd Canadian Infantry Brigade forward with the 5th (Western Cavalry) and 8th (90th Rifles of Winnipeg) battalions holding the front line, the four companies of the 7th (1st British Columbia) Battalion dispersed in brigade support, and the 10th (Alberta) Battalion as the brigade reserve. He placed the 1st Canadian Infantry Brigade behind them as the divisional reserve, and assigned the 3rd Canadian Infantry Brigade to Corps reserve.

THE GERMAN ATTACK

German commanders understood well the operational and tactical value of the Ypres salient as well as the threat posed to them by the Canadian position that straddled its high ground. If they could capture Tor Top, the hills around it, and the positions beyond along Observatory Ridge, the Germans would gain vital ground to the rear of the Canadian position while at the same time robbing them of direct observation of their own rear areas. Once consolidated, the Germans would be able to bring both indirect and direct fire to bear on the Canadian support lines, and with enough pressure perhaps even force the Canadians to withdraw completely out of the salient. Even if not successful in taking the position in an all-out attack, the Germans felt confident that the tactical value of the salient to the Entente would force them to drain resources from their preparations at the Somme to commit to the defence of the Ypres salient. It was a calculated risk certainly, but in the eyes of the Imperial German Army, one that was certainly worth taking.

The German Fourth Army held the line along the Ypres salient opposite the Canadians. In April 1916, Its commander, the Crown Prince Rupprecht, gave the task of seizing Mount Sorrel to General der Infanterie Freiherr Theodor von Watter, commander of *XIII. (Königlich Württembergisches) Armeekorps.*[7] Overall, Watter's Corps were in good fighting condition despite increasing food, material, and manpower shortages, and were prepared to undertake any new offensive operations as ordered. Watter's Armeekorps at the time consisted of two infantry divisions, 26 Division (*1. Königlich Württembergische)* under the command of General der Infanterie Duke Wilhelm von Ulrach[8], and 27 Division *(2. Königlich Württembergische)* under the command of General der Infanterie Friedrich Woldemar Franz Graf von Pfeil und Klein-Ellguth.[9] In addition to his infantry, General Watter's Württemberg Corps included one battalion of the 13th Foot Artillery Regiment as well as a number of pioneers, combat support, aviation, and other ancillary troops.[10]

Once ordered to take Mount Sorrel, von Watter began his planning and preparations in earnest. His preliminary activities were well concealed, but the Canadians were able to discern from their own reconnaissance and patrolling that it was only a matter of time before the enemy attempted an assault. On the morning of 2 June 1916, Watter's XIII Armeekorps ended Canadian speculation and launched its much-anticipated attack. Though the exact moment of the initial assault came as a surprise to the Canadian defenders and caught the commanders of both the 3rd Canadian Division and the 8th Canadian Infantry Brigade out in the initial bombardment, there was unfortunately little more the Canadians could do to mitigate the risks presented to them.[11] The ferocity of the initial German assault was successful in crippling the leadership and communications of the defenders as well as seriously disrupting their chain of command. The GOC of 3rd Canadian Division, Major-General Malcolm S. Mercer, was first wounded by the German bombardment and then later killed by friendly fire. The GOC of the 8th Canadian Infantry Brigade, Brigadier-General Victor Williams, was also wounded and later made a prisoner. The commanding officers of several battalions were likewise either wounded or killed, and for several hours the Canadian defence of Mount Sorrel rested on the shoulders of brave captains and sergeants as the higher headquarters were paralyzed.

Despite putting in a terrible four-hour preparatory bombardment against the Canadian positions, the Germans were undoubtedly surprised when those who managed to survive this initial punishment continued to fight. Weather, ground, artillery, and logistics support all favoured the attacker, but the 3rd Canadian Division managed to contest every piece of ground as well as inflict serious casualties on the leading waves of the assaulting Württemberg corps. Far from being a walkover as more popular histories have suggested, the Canadians showed great tenacity and willpower, requiring the Germans to resort to flamethrowers and fierce hand-to-hand combat to subdue the last pockets of resistance along the salient. Even then these efforts did not succeed in completely routing the 3rd Canadian Division off the high ground, and it would take all day to finally secure the majority of their objectives.[12]

Positioned on the southern flank of the 3rd Canadian Division, the 1st Canadian Division commanded by Arthur Currie was subjected only to demonstrations and diversionary attacks by the German 117th Division on 2 June in order to fix the formation in place and deny its support to the friendly units on its left. The German feint was not successful, for the 3rd Canadian Infantry Brigade put an end to the enemy nuisance at its front without much difficulty and was then able to turn its machine guns into the exposed flanks of the Württemberg's 120th Regiment that was at the time swarming over the crest of Mount Sorrel. The GOC 3rd Canadian Infantry Brigade, Brigadier-General George Stuart Tuxford, employed his forward infantry battalions, the 5th and 7th, with great effect and did much to blunt the enemy's penetration into the Canadian defensive positions immediately to his north.[13]

By the evening of 2 June the Württemberg attack had stalled and initial consolidation began. Though General der Infanterie von Watter had achieved both surprise and some success in taking parts of the salient, it had cost his XIII Corps terribly. Worse, von Watter was now stuck defending the same ground that the Canadians found so difficult to hold, and it was certainly made even more untenable by the day's battle damage. Still, the situation on the ground was very confused at the moment, especially for the Canadian defenders whose own front line was in disarray. It was unknown to them whether the Germans controlled the entire salient or only parts of it. The whereabouts of both the division and brigade commander remained unknown, the fate of several units and their commanders were unknown, and perhaps worst of all it was not known whether the German attack had finished or if this initial penetration was the beginning of something much larger. Lieutenant-General Sir Julian Byng was therefore forced to take action immediately, employing whatever countermoves force he could cobble together before the Germans had a chance to prepare proper defences against any Canadian counter-attacks or even execute further assaults of their own towards the town of Ypres itself.

THE HASTY COUNTER-ATTACK AND ITS PROBLEMS

The decision to immediately counter-attack and retake lost ground knowing full well that support was limited and that casualties might be high may offend modern sensibilities, but at the time it was considered an appropriate unquestioned and unsentimental approach to combat operations. Corps Commanders were not selected for their meekness in battle, and in 1916 generals were still routinely "degummed" if they failed to achieve success in combat operations.[14] The threat of being cashiered mixed with the natural desire to come to grips with the enemy was more than enough encouragement to be aggressive on the battlefield. Lieutenant-General Sir Julian Byng, appointed to command of the Canadian Corps only a few days before the German assault, very likely felt even greater pressure in his new post to act quickly and get results.[15]

Arthur Currie had spent most of the morning on 2 June 1916 watching the battle unfold on his left, and knew that his division would soon be called upon to counter the enemy penetration along the salient. When sounds of intense rifle and machine gun fire were reported coming from Mount Sorrel at 1:15 pm, Byng ordered Currie to have his 2nd Canadian Infantry Brigade be prepared to retake the high ground if it fell into enemy hands.[16] Currie immediately sent orders along to its commander, Brigadier-General L.J. Lipsett, who returned his acknowledgement of the order with the additional information that the trenches to his front were now definitely occupied by the enemy. Lipsett also added that his forward commanders could observe German soldiers swarming over the top of Mount Sorrel and pushing on through Armagh Wood.[17]

No further instructions were received by Currie from Corps Headquarters (HQ) until preliminary orders to retake Observatory Ridge arrived at his headquarters at 4:15 pm. Brigadier-General Tuxford was with Currie when the initial Corps order arrived, so the division commander was able to get his 3rd Canadian Infantry Brigade moving almost right away. Word from the 3rd Canadian Division HQ finally got through to Currie at 4:30 pm confirming some of the German gains, and they also delivered the grim news that the division was essentially leaderless, severely battered, and unable to mount any sort of counter-attack.

Then there were further complications. Instead of being ordered to lead his own division in the counter-attack, Currie received notice fifteen minutes later that Brigadier-General E.S. Hoare Nairne, GOC of the Lahore Divisional Artillery who had been given temporary command of the 3rd Canadian Division in Mercer's absence, would coordinate and issue all orders for the Canadian counter-attack against Mount Sorrel. Furthermore, the 1st Canadian Division was ordered to

SECRET

1st Canadian Division
Operation Order No.73
June 2nd 1916

1. Latest reports show the enemy to be in possession of MOUNT SORREL and part of ARMAGH WOOD, the Eastern portion of OBSERVATORY RIDGE, including our Front Line Trenches from Trench 53 to Trench 62 and to be consolidating.

2. All ground lost today will be retaken tonight under the orders of Brigadier General NAIRNE, commanding 3rd Canadian Division.

3. For the above purpose the 3rd Canadian Infantry Brigade reverts from Corps Reserve and is placed at the disposal of the 3rd Canadian Division.

4. The 2nd Canadian Infantry Brigade will also receive orders from the 3rd Canadian Division for the retaking of MOUNT SORREL.

5. One Battalion, 1st Canadian Infantry Brigade, now in occupation of G.H.Q. 2nd Line, will be placed under the orders of the G.O.C. 2nd Canadian Infantry Brigade. If this battalion is moved, Divisional Headquarters will be informed and it will be replaced by another battalion from DICKEBUSCH HUTS, of the 1st Canadian Infantry Brigade.

Signed,
R.H. Kearsley,
Lieutenant Colonel
General Staff
1st Canadian Division

Issued at 10.40 p.m.

FIGURE 1.1 – 1ST CANADIAN DIVISION OPERATION ORDER NO.73

detach two of its three brigades under operational command to Hoare Nairne for his employment. Currie tactfully and diplomatically protested the decision to break up his division to Byng, but the new Corps Commander would not be swayed. Byng felt it would not be possible to relieve the battered 3rd Canadian Division given that it was still in contact with the enemy; therefore the only option was to immediately reinforce it with other units. Still, Currie expressed concern about trying to reorganize divisions in the dark – there was too little time to cover the distances, organize and coordinate the composite formations, re-establish communications, assess the enemy situation, issue orders, and prepare for battle. Byng ignored these issues and simply told both Currie and Hoare Nairne to get on with it. Seeing no other out Currie conceded the point to his superior and left to see what could be done elsewhere to mitigate the risks. Lieutenant-General Byng's decision was confirmed by Currie's HQ at 10:15 pm with the arrival of Canadian Corps Operation Order No.17.[18] From this order, Major-General Currie issued his own orders (Figure 1.1) just a half hour later.

Today, commanders are routinely taught a simple formula when preparing for a battle. Essentially, to ensure that subordinates have sufficient time to make their own preparations, the commander determines how much time there is in total between the point at which he or she receives their orders and when the attack is to begin, and then allots one third of this time for their preparation and gives two thirds of this time to subordinates for their own preparation. Therefore if the commander received his orders at noon for an attack at 9:00 pm that evening, he should expect to deliver his own orders to his subordinate commanders no later than 3:00 pm. However, any number of factors can impose arbitrary constraints on this formula, and seldom do these constraints and restraints serve to the benefit of the planners.

Figure 1.2 illustrates Major-General Currie's battle procedure for the initial counter-attack on 3 June. As can be seen, the constraints and restraints faced by Currie and his staff to detach their brigades to 3rd Canadian Division were formidable, with much arbitrary timing forced on his headquarters and little support to ease the compression of his battle procedure. Though the battle was well underway by the mid morning of 2 June, it took nearly four hours for Corps HQ to issue Currie his first warning order. He received no further orders until 4:15 pm, and wasn't given any clear indications on how his division was to be employed in the counter-attack until after 8:00 pm, after which he was forced to wait another two hours until Corps HQ issued its own operation order to Currie.

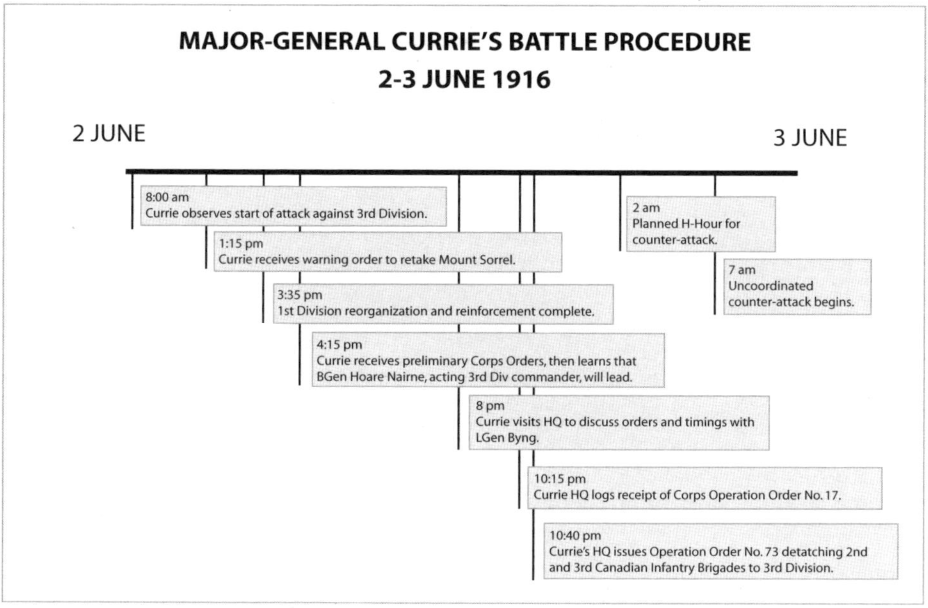

FIGURE 1.2 – MAJOR-GENERAL CURRIE'S BATTLE PROCEDURE

Despite only receiving his formal orders from Corps HQ at 10:15 pm, Currie's brigades were expected to reorganize, move, reorganize again, prepare for battle, secure their lines of departure, and launch their counter-attack in less than three hours. That he was allowed so little time to detach and move his brigades under command of another division for a complex counter-attack employing composite organizations is revealing – either Byng and Hoare Nairne did not appreciate the tactical and logistical problem they faced or they were willing to take considerable risk and launch a hasty counter-attack before the Germans could consolidate their newly won positions. Given that both men were experienced soldiers it is more likely that fear and concern for the strategic security of Ypres drove their actions more than lack of appreciation. Still, there is no adequate explanation why Byng did not employ Currie or perhaps another senior Canadian Corps officer to organize and lead the counter-attack instead of a British brigadier from the divisional artillery. Byng was already accepting considerable risk in attempting to recapture Mount Sorrel as quickly as possible, therefore it would have made more sense to give the task to someone intimately familiar with Canadian units and chain of command.

The devil is always in the details, and General Byng's attempt to rush his formation through its battle procedure had its consequences. Manoeuvring a composite division over unfamiliar ground in the dark while being constantly harassed by enemy fire could only happen so fast. The battlefield was a mess of destroyed material, dead men, and the dying. Trenches had caved in from the shelling, and for units that had not previously occupied this sector of the front nothing appeared familiar in the darkness. As of June 1916, official British Army maps issued down to formations still only showed enemy trenches, which obviously proved of little value when trying to navigate the friendly front line.[19]

Securing the lines of departure for the counter-attack also took time, as did ensuring that all units involved were properly briefed on the signals and communications that would initiate the assault. It was decided to fire six green rockets simultaneously to begin the attack, but even this simple method of signaling was not without its own problems, as ammunition could prove unreliable.

It is little wonder then that the attack was repeatedly delayed as units attempted to clear their front lines and get organized for battle. The many difficulties experienced took their toll on overall timings, and instead of assaulting under the cover of darkness; the Canadian counter-attack did not begin until after 7:00 am the next morning. As might be expected, disaster struck immediately. Several of the rockets employed to signal the attack misfired, and it eventually took fourteen rockets to get six to illuminate the sky. The units waiting for the signal became confused over whether the signal they observed was legitimate or not. Some battalions launched into their attack but others waited for further confirmation. This poor coordination and division of units resulted in both the forfeiture of surprise as well as any concentration of force against the objective.

Poor planning and coordination ensured that Brigadier-General Hoare Nairne's hasty counter-attack plan did not survive first contact with the enemy. Currie's 7th Battalion was employed on the right flank of Hoare Nairne's attack, supported by the 10th Battalion, while his 14th and 15th Battalions attacked on the left with the 13th Battalion in support. What little artillery support there was for the attacking infantry proved to be of no help, as Byng and Hoare Nairne had not allowed any time to determine the exact location of the new German front line. This allowed the German machine gunners, especially, to wreak havoc on the uncoordinated and unsupported Canadian infantry.

After suffering considerable casualties for only limited gains, General Byng decided to halt the counter-attack. He had just learned the hard way that despite being exhausted these were no ordinary German troops, and that if he was to unhinge them from the salient he needed a solid plan as well as a robust leader who could execute it. Without further delay, Byng opted to put his faith in Major-General Arthur Currie.

DELIBERATE ATTACK AND SUCCESS

After the failure of the 3 June attack, Major-General Currie was assigned to succeed where Brigadier General Hoare Nairne had failed. Given that his orders were to execute a deliberate attack to retake Mount Sorrel, Currie insisted on being given sufficient time to conduct proper battle procedure so that he could have a good chance of success. Byng granted Currie's request, and left the commander of 1st Canadian Division to set about laying the groundwork for what would become the Canadian Corps' first set-piece attack.

Currie knew that to be successful at Mount Sorrel, he had to shape the conditions of the battlefield to his advantage. The Württemberg units were rapidly consolidating their newly captured positions, putting out new wire obstacles in front of their trenches as well as through Armagh wood.[20] Canadian forward sentries also noted the sudden appearance of several periscopes along their old trenches, which was followed soon after by severe enemy shelling along all the lines of communication in the Canadian rear area. Currie needed to quickly deny General von Watter this ability to directly observe and shell Canadian positions indiscriminately and instead turn the Canadian artillery onto the German positions, but to do it he needed actionable intelligence on his enemy.

Over the next couple of nights Major-General Currie pushed out a massive reconnaissance effort to accurately measure the exact state of German defences and capabilities. This tremendous intelligence collection operation included the identification of enemy units opposite them, the locating of new enemy trenches and the updating of Canadian maps, and the compilation of enemy target lists.

Every aspect of the new German defence was identified – assembly areas, dugouts, ammunition and supply dumps, routes of approach, command posts, observations posts, machine gun emplacements, artillery gun emplacements, and communications posts.[21] All enemy locations were verified against maps, ground patrol reports, enemy interrogations, and aerial photographs. By the evening of 6 June, Major-General Currie had a highly developed target list to issue to both the Canadian Corps Heavy Artillery as well as the 1st Canadian Divisional Artillery. These formations began a deliberate bombardment against these targets on the morning of 7 June.

Yet even as Currie began to gain the upper hand at Mount Sorrel, the Germans attempted to steal the initiative of the battle away from the Canadians. On the evening of 5 June, XIII Armeekorps attempted to advance near Maple Copse once again, but the Canadian artillery and infantry were this time able to beat them off without any noticeable gain. At approximately noon on 6 June, however, the Germans struck again in force, this time to the north at Hooge. The 6th Canadian Infantry Brigade had just occupied the position the night previous having relieved the remnants of 7 CIB, and its forward battalions received the full brunt of an artillery bombardment similar in intensity to that launched on 2 June. The 28th (Northwest) and 31st (Alberta) Canadian Infantry Battalions were hammered by the shelling all afternoon, but the final insult was delivered at 3:00 pm when four mines were detonated underneath the Canadian front trenches. The blast annihilated a whole company of 28th Battalion soldiers in an instant, and when the Germans swarmed into the gap there was no one left to fight them.[22] The German assault at Hooge was eventually stemmed by the RCR and the 31st (Alberta) Canadian Infantry Battalion, but not before the village farm and its surroundings were captured.

Having had about enough of it, Currie issued Operation Order No.75 on 6 June at 11:55 pm, explaining the plan for the assault to recapture Mount Sorrel.[23] Though the plan did not incorporate a sophisticated rolling barrage or the same level of counter-battery work as observed in his later set-piece battles, it is evident that his penchant for meticulous intelligence preparation of the battlefield followed by detailed mission analysis and orders preparation had asserted itself. Currie considered every detail of the assault right down to the individual soldier loads, ensuring that the men carried enough water with them as well as other necessary items such as shovels and sandbags to quickly rebuild destroyed defences. Most important, he ensured that sufficient fire support arrived with the infantry on the objective, so that the soldiers could hold their trenches against enemy counter-attacks that were sure to follow. Unfortunately, however, the weather turned sour and Currie was forced to cancel the order the following morning. He would have to wait a few days before getting his chance to strike.

Despite being forced to delay the assault, Currie still had his brigades aggressively patrol no man's land to keep his intelligence assessments of the enemy up to date. Aerial reconnaissance of the German lines was carried out on 7 June, followed by night patrols on the night of 7-8 June. During one outing, a small scout detachment from the 10th Battalion thoroughly searched parts of the 3rd Canadian Division's old trenches where it ambushed an enemy patrol killing one soldier. The German rifleman was later identified as having belonged to the 4th Company, 1st Battalion, 11th Reserve Infantry Regiment, the unit that had recently replaced the 120th Regiment in this part of the front line.[24]

Constant pounding of the German lines by Canadian artillery throughout 8 and 9 June made life increasingly miserable for the new defenders of Mount Sorrel. Shelling of the trenches around Hill 60 and the Snout damaged them beyond any use, and the Germans abandoned a number of other conspicuous targets including Armagh House. Further reconnaissance of the German positions throughout the day noted that their wire obstacles were severely damaged and failing. One Canadian report noted, "There is a single strand of barbed wire 20 yards in front of the enemy parapet, from I.30.a.3.2 to I.30.a.5.6. In the same neighbourhood, and ten yards from the enemy's parapet, there is some loop wire not securely staked to the ground."[25]

With the German defences methodically reduced by Canadian artillery, Major-General Currie was able to pry the initiative of the battle away from General von Watter. Having conducted a proper intelligence and mission estimate in order to complete a properly detailed plan, all the Canadians needed now was reasonably good weather, an element of surprise, and some good luck. And just as von Watter received the advantage on 1 June, so too did Currie get the advantage on the evening of 12 June.

For the counter-attack, Currie's trusted subordinates, Brigadier-Generals Lipsett and Tuxford, each commanded one of the composite brigades. Lipsett's brigade, composed of the 1st, 3rd, 7th and 8th Canadian Infantry Battalions, was to attack on the right and recapture Mount Sorrel. In the centre and on the left, meanwhile, Tuxford's composite brigade composed of the 2nd, 4th, 13th, and 16th Battalions would make an attack against Tor Top. As they formed up for the attack on 12 June, Canadian artillery pounded the German defences between Hill 60 and Sanctuary Wood for ten hours, focusing in particular on the high ground around Hill 60 and the snout. The gunners, in receipt of good intelligence and having had time to prepare, were extremely accurate and effective in suppressing the German positions while the Canadian infantry moved out into no man's land and away from the expected German counter barrage zones.

The Canadian composite brigades launched their counter-attack against Mount Sorrel at 1:30 am on 13 June. Assaulting up the ridge in a driving rainstorm and covered on both flanks by well-placed smoke barrages, the Canadians stole into the German front lines before any effective defence could be mounted, killing the resistors and capturing the rest. In all, 191 German prisoners were taken, and despite the terrible conditions of the ground, Brigadiers Lipsett and Tuxford were able to consolidate their positions quickly. As well, because Currie had planned ahead for his artillery to keep supporting the infantry once the heights were retaken, the Canadians were able to repulse without difficulty a German counter-attack against Mount Sorrel at 6:45 am on 13 June. The enemy tried a second time at 9:00 am, but when this too was defeated, the Germans declared themselves beaten and made no further attempts against Mount Sorrel that day.

THE VERDICT

All things considered, the aggressive application of leadership ensured that the Canadian Corps fared better at Mount Sorrel than it might have otherwise. British preparations for the July offensive on the Somme precluded any requests for additional resources or artillery to defend the Ypres sector. The weather throughout late May 1916 favoured the attacker and concealed German preparations. It also hindered the Canadian ability to prepare detailed defensive fires or air assisted counter-battery programs. The ground itself was notoriously difficult to defend, and by mid-1916 was a mess of blasted tree roots and high water tables. This made it a challenge to effectively dig into, and provided little natural protection against concentrated enemy artillery.

The 3rd Canadian Division had made what preparations it could to defend against an attack it knew was imminent, but predicting the exact moment of attack was a difficult if not altogether impossible task even under the best circumstances. The Canadians aggressively patrolled no man's land and learned what they could. The combat intelligence that was gained up to the night of 1 June could not ascertain the locations of new German artillery or staging positions. No new enemy troops were observed in opposite the Canadian sector. Nothing implied an attack would take place the following morning. The Canadians had not failed to properly prepare for the defence of Mont Sorrel as some historians have suggested, the conditions did not favour them and they were simply beaten by a well-prepared and better-supported adversary who had ground, time, weather, and luck on their side.

Yet despite these conditions, Major-General Arthur Currie not only proved himself a talented tactician, but also a good learner. A veteran of all Canadian battles in 1915, he was able to successfully capture relevant observations and experiences from each engagement he fought and institutionalize these "best practices" in his next operation. This is not to suggest that Currie did not make mistakes later in the war, like any general he was certainly fallible. Yet when compared to his Division and Corps level peers, Currie continued to stand out as one who grasped the complexities of modern war, and as one who was capable of mastering both the division and corps level tactical battle.

Written some years later in 1932, the British official history of the battle of Mount Sorrel noted that "the first Canadian deliberately planned attack in any force had resulted in an unqualified success."[26] Given some of the less flattering attitudes displayed by the British official historian towards Canadian operations and in particular Sir Arthur Currie's war record, this was undoubtedly considered a rare compliment by all towards the man who led the Canadian Corps to victory.

1 There are few exceptions to this general trend. For examples see Patrick Brennan, "From Amateur to Professional: The Experience of Brigadier General William Antrobus Griesbach," in Briton Bush ed. *Canada and the Great War: Western Front Association Papers.* (Montreal: McGill-Queen's University Press, 2003), pp.78-92; also Reginald H. Roy. *For Most Conspicuous Bravery: A Biography of Major-General George R. Pearkes VC Through Two World Wars.* (Vancouver: UBC Press, 1977).

2 Daniel G. Dancocks. *Sir Arthur Currie: A Biography.* (Toronto: Methuen Books, 1985).

3 Historians have also made the excuse that the lack of detail in Currie's own journals and professional papers during this period precludes us from deriving a better understanding of him during the earlier years of the war.

4 Ian M. Brown. "Not Glamorous, But Effective: The Canadian Corps and the Set-Piece Attack, 1917-1918", *The Journal of Military History*, 58 (July 1994): 421-444.

5 Peter Barton. *The Battlefields of the First World War.* (London: Constable and Robinson, Ltd., and the Imperial War Museum, 2005), pp.118-119. This book and its CDs offer amazing panoramic landscape shots of the Mount Sorrel and Hooge battlefields during 1915 and 1916.

6 Interestingly, the name "Mount Sorrel" is not an indigenous one, but rather the name given to this high feature by a British officer from the Leicestershire Regiment who before the war served as the Director of the Mountsorrel Granite Company.

7 Watter's career is detailed in Hans Möller-Witten. *Geschichte der Ritter des Ordens „pour le mérite" im Weltkrieg.* (1935).

8 Born as Prince Wilhelm Karl Florestan Gero Crescentius of Urach, Count of Württemberg, he was the elder son of Wilhelm, 1st Duke of Urach, the head of a morganatic branch of the Royal House of Württemberg, and his second wife, Princess Florestine of Monaco, occasional acting Regent of Monaco. After the war he made a failed attempt to be anointed the King of Lithuania.

9 The young Leutnant Erwin Rommel, destined to become one of the most famous German generals in the Second World War, originally served as an officer with No. 7 Company, Infanterie-Regiment König Wilhelm I (6. Württembergisches) Nr. 124 of the 53. Kgl. Württembergische Infanterie-Brigade. (27 Div). However, he was transferred to the Württemberg Mountain Battalion in the fall of 1915, and was fighting on the Eastern front at the time of the battle of Mount Sorrel.

10 Günter Wegner, *Stellenbesetzung der deutschen Heere 1815-1939.* (Biblio Verlag, Osnabrück, 1993), Bd. 1

11 Library and Archives Canada (LAC). Record Group (RG) 9 Series III-D-5. Canadian Corps War Diary (CCWD). Intelligence summaries clearly show that the 3rd Canadian Division knew an attack was imminent but due to a number of factors were unable to ascertain the exact timing of the assault.

12 The Württembergers themselves were so impressed by the Canadian defence of the salient that they received special mention in several postwar Württemberg histories of the battle.

13 The 5th Battalion was subjected to a heavy bombardment throughout 2 June but was able to effectively engage the enemy despite the shelling.

14 "Degummed" was a colloquialism for being relieved of command duties.

15 For a brief discussion of British command philosophy see Simon Robbins. *British Generalship on the Western Front.* (London: Routledge, 2005).

16 LAC. RG 9 Militia and Defence, Series III-D-3 Vol.4828 Reel T-7188, File 53. 1st Canadian Division Summary of Operations dated 8 June 1916, p.1.

17 Ibid., p.1.

18 The 1962 official CEF history states that Byng issued his orders for the counter-attack at 8:45 pm, but the 1st Canadian Division communications log kept during the battle records the orders as received at 10:15 pm. Even if borne by a lost runner, it would not have taken over an hour to deliver the orders to 1st Canadian Division HQ. Given the other factual errors in the official account of the battle it is likely that the 8:45 pm timing is incorrect.

19 Peter Chasseaud. *Rats Alley: Trench Names of the Western Front.* (Gloucestershire: Spellmount, 2006), pp.83-85.

20 LAC. RG 9 Militia and Defence, Series III-D-3 Vol.4828, File 53. Summary of Intelligence, 1st Canadian Division, 5 June 1916.

21 Dan Jenkins. "Winning Trench Warfare: Battlefield Intelligence in the Canadian Corps, 1914-1918". (Ottawa: Carleton University Unpublished PhD dissertation, 1999), p.189.

22 A Company, 28th Battalion, occupied trenches 70 to 72 where the mines were blown. With the exception of a few men dispersed to other tasks, the entire company was effectively annihilated in a single stroke.

23 LAC. RG 9 Militia and Defence, Series III-D-3 Vol.4828, File 53. 1st Canadian Division Operation Order No.75, dated 6 June 1916.

24 LAC. RG 9 Militia and Defence, Series III-D-3 Vol.4828, File 53. Summary of Intelligence, 1st Canadian Division, dated 8 June 1916.

25 Ibid.

26 Brigadier-General Sir James Edmonds. *History of the Great War: Military Operations – France and Belgium, 1916.* (London: Macmillan and Co. Ltd, 1932), p.241.

CHAPTER 2

"A Leap in the Dark" – Intelligence and the Struggle for the St. Eloi Craters

Reassessing the Role of Major-General Richard Turner

DAVID CAMPBELL

Military leaders base their decisions upon the best possible intelligence regarding the dispositions and material capabilities of their own forces and those of the enemy. If a commander's subordinates or superiors supply him with faulty intelligence, then he is left in the predicament of having to formulate plans based upon inaccurate information that, at the time, appeared to be correct. The commander may not have directly generated such false intelligence, but he is responsible for interpreting and acting upon it. The Prussian military theorist Carl von Clausewitz, appreciating the pitfalls presented by battlefield intelligence, once remarked that:

> Many intelligence reports in war are contradictory; even more are false, and most are uncertain. What one can reasonably ask of an officer is that he should possess a standard of judgment, which he can gain only from knowledge of men and affairs and from common sense. He should be guided by the laws of probability. These are difficult enough to apply when plans are drafted in an office, far from the sphere of action; the task becomes infinitely harder in the thick of fighting itself, with reports streaming in. At such times one is lucky if their contradictions cancel each other out, and leave a kind of balance to be critically assessed.[1]

Such was not Major-General Richard Turner's luck during the battle for the St. Eloi craters in April 1916. The battle is remembered as one of Canada's worst single defeats of the First World War. It was the first major action of Turner's 2nd Canadian Division and it proved to be a bloody baptism. Between 3 and 16 April, the division suffered 1,373 casualties and lost a substantial amount of ground that had been gained earlier by British troops whom the Canadians had relieved.[2] In the aftermath of this failure, a number of Canadian and British officers were removed or reassigned, the most notable of whom was Turner's superior, the Canadian Corps commander, Lieutenant-General Sir Edwin Alderson.[3] The fiasco also represented another blot upon the record of Turner himself, who had performed poorly as a brigade commander during the Second Battle of Ypres in April and May of 1915.

Before the 1990s, published explanations of the defeat at St. Eloi tended to focus upon the intractable obstacles faced by Turner's division.[4] These obstacles included poor weather, terrible ground conditions, intense bombardments from the German artillery, communications disruptions, dud artillery shells, cantankerous Ross rifles, and the relative inexperience of 2nd Division's troops. On the other hand, later historical accounts have been increasingly critical of the performance of Turner, his staff, and some of his subordinate commanders. These criticisms focus on Turner's inability to recognize that there were serious problems with the disposition of his troops – problems that played a large part in 2nd Canadian Division's defeat at St. Eloi. Turner's failure to identify and rectify these errors has drawn heavy fire from some historians, who have laid the lion's share of responsibility for the disaster at his feet.[5]

There are, however, other approaches that focus not so much on the decision-makers themselves, as on the information upon which their decisions were based. If we examine in detail the intelligence that was generated during this battle by British and Canadian troops, we can put Turner's decisions in their context and better understand why he failed to realize that many of his troops were occupying the wrong positions. This approach leads to the conclusion that Turner faced a challenge at St. Eloi which may have been insurmountable, owing in part to the provision of inaccurate front-line intelligence.

The little hamlet of St. Eloi, situated some five kilometres south of Ypres, was faced by a small enemy-held salient that had been formed by a German attack in March of 1915. The base of this German salient was approximately 600 yards wide and it extended 100 yards north into the lines held by elements of British Second Army. In fact, the lines in this sector ran east-west instead of in the usual north-south configuration that typified much of the western front between Nieuport and Noyon. But the most significant feature of this rather miniscule piece of real estate was a feature known as "the Mound," a slight elevation of some ten to twenty feet above the surrounding boggy terrain. From their observation posts on this diminutive knoll, the Germans overlooked much of the immediate area, and even though British shells had consistently plastered the Mound for over a year, the Germans continued to make good use of it. For General Herbert Plumer, GOC of Second Army, this was a feature worth denying to the enemy, and one that would prove useful to British observers if successfully taken.

Another motive that Plumer had for attacking at St. Eloi was his desire to avenge the loss of a similar bit of high ground to the east of the St. Eloi salient. The Germans had captured British positions at "the Bluff" on 14 February, and when immediate British counter-attacks failed to reclaim the lost ground, Plumer decided to retaliate at a different point in his sector. He chose the German salient near St. Eloi as his target. However, there were problems with Plumer's plan and his choice of target. For one thing, even if the British managed to eliminate the German salient and capture the Mound, the entire area was still under effective German observation from the commanding heights of the Wytschaete Ridge, which lay further to the south. In addition, the battlefield around St. Eloi was a muddy wasteland that would be difficult to navigate. Innumerable shell holes and old mine craters of varying size and depth pockmarked the landscape in the vicinity of the Mound. Nevertheless, despite the dismal nature of the terrain, Plumer and his staff hoped that the coming spring weather would dry the ground enough to make operations viable. Unfortunately for the troops who would have to carry out the scheme, these hopes would prove delusive.[6]

Plumer, a major exponent of mine warfare, decided to take the German positions at St. Eloi by tunnelling under their lines and laying six huge mines. When detonated, the mines would simply demolish the German trenches from below, thereby allowing assaulting units from Second Army to rush in and occupy the ground. Mining had been going on at St. Eloi long before the loss of the Bluff.[7]

Since August 1915, the British had sunk three large shafts and driven them steadily forward. By March 1916, they had galleries excavated under the German trenches. The four central and largest mines were laid directly beneath the German lines, including the Mound itself. The remaining two mines, on the extreme left and right, fell short of the German front line, owing to enemy countermining which had prevented their extension. One is left wondering why, if he desired to secure the Mound as an observation point for his own men, Plumer planned to blow much of it away with these mines.[8]

The Canadian Corps, which included the 2nd Canadian Division, was part of Plumer's Second Army. But the Canadians were not tasked with the initial assault of the German positions following the blowing of the mines. That job would fall to the 3rd Division of the British V Corps. The Canadians were, however, ordered to prepare to relieve the British assault troops once the new positions had been consolidated. It would be left to the Canadians to hold the newly-won ground, and face the expected German counter-attack.

The battle itself commenced at 4:15 am on 27 March 1916 with the detonation of the six mines, the four largest of which obliterated the German front lines, leaving fresh, gaping craters. The craters resulting from this cataclysm were designated from west to east as Craters 1, 2, 3, 4, 5, and 6. Battalions of the 9th Infantry Brigade of the 3rd British Division quickly moved to the assault, even as the last clods of earth were plummeting to the ground. The battalions on the right flank of the British attack succeeded in reaching their objectives in less than half an hour and with comparatively little difficulty, owing to the shock of surprise that the titanic explosions wrought upon the German front line defenders.[9] Unfortunately for the attackers on the left flank, enemy machine gun fire took a heavy toll, and progress toward the ultimate objective, the German second line, was checked. In these initial actions, the British managed to capture Craters 1, 2, 3, 6 and an older crater later numbered as 7. But a gap existed in their new front line. This had been caused by the hold-up on the left flank, and the vigilant Germans exploited the confusion, infiltrated through the gap, and occupied Craters 4 and 5.

In fact, it was initially reported by the 4th Royal Fusiliers, the battalion that had been stalled on the left flank, that they managed to reach German positions south of Craters 4 and 5. Subsequent events proved this report wrong, and the position actually reached by the Royal Fusiliers was in front of Craters 6 and 7. The apparent reason for this mistake, according to the British Official Historian, was the "changing landscape" of the crater sector.[10] The blowing of the mines had totally altered the topography of the German salient, rendering old maps essentially useless. British and German artillery barrages continued to churn up and alter the ground, and, to make matters worse, the blowing of the mines had completely disrupted the drainage system in the area, creating a muddy moonscape of waterlogged shellholes, craters, and trenches. The confusion experienced by British troops in the bewildering quagmire of the crater sector would, unfortunately, be shared by the Canadians who were assigned to relieve them.

A combination of ground and aerial reconnaissance eventually revealed the truth about actual German dispositions in Craters 4 and 5.[11] After several days of hard fighting, during which the 3rd British Division was forced to throw the Divisional Grenade School staff into the fray, the entire crater sector was finally secured by daylight on 3 April. To ensure that there was no confusion about the positions occupied, the GOC of 3rd British Division, Major-General Aylmer Haldane, made a personal reconnaissance along the entire length of the new front line.[12] The British were by now completely exhausted, and consequently had to be relieved by the Canadians three days ahead of schedule. The original scheme had called for the relief to occur on the night of 6-7 April.[13]

The 6th Infantry Brigade of Major-General Richard Turner's 2nd Canadian Division relieved the beleaguered British troops and took over the newly-won positions at St. Eloi on the night of 3-4 April 1916. The Canadians were thrust into a situation that was less than ideal. Writing after the war, Canadian Official Historian, Colonel A.F. Duguid, declared that:

> In all the history of war there can rarely have been a more disadvantageous position than the 1000 yards of new front line taken over by the 27th Battalion of the 6th Canadian Infantry Brigade at 3 a.m. 4th April. The whole area lay spread out like a map under the eyes of the Germans in their observation posts on the Wytschaete ridge, with the sun at their backs. For British artillery observers there was no such commanding

> position; the lips of the craters formed the sky line for the 500 most critical yards and only when the sun was low and the shadows revealed conformation was it possible to identify each crater by sight as well as by bearings; the front line lay invisible in the dead ground 150 yards beyond.[14]

Still worse was the fact that the new reserve line, which had been the British front line on 27 March, was completely destroyed by enemy shelling and virtually untenable for a thousand yards. The entire area was still as much a pond-dotted morass as when the British had held it. Many trenches were submerged in two to three feet of water and still choked with dead and wounded British and German soldiers.[15] In his published account of the operations at St. Eloi, Lord Beaverbrook painted a vivid picture of the miserable terrain endured by troops on the battlefield, noting that, "As men splashed from shell-hole to shell-hole – and the surface of the earth consisted of nothing else – they sank up to the armpits and could find no grip for their feet. One of the strongest men in the 2nd Division has declared that after sixty yards of this work he was incapable of going further."[16] According to D. E. Macintyre, Staff Captain and Intelligence Officer for the 6th Brigade, "Major-General Turner, worried about the situation, made a personal reconnaissance of the whole line [on 4 April] and was, I believe, the only man who ever got completely around. There never was another opportunity...the spectacle of a Major-General making such a thorough and prompt inspection of a new and very active sector was a novel one even to the troops..."[17] When Turner visited Macintyre's dugout early that morning, he admitted that the situation was much worse than he had expected.

There was no continuous front line at all – only a series of disconnected slit trenches, old German fire trenches (which faced the wrong way), deep, narrow drains, and sodden shell holes just south of the craters. But in Turner's opinion, the worst feature was the lack of any viable communication trench through the centre of the crater sector, which was needed to link the new forward positions with those to the rear. Instead, movement had to be conducted around the flanks of the craters, and the four central craters (Nos. 2, 3, 4, and 5) were blown so close together that they presented a virtually impassable barrier to direct movement between the front and the rear.[18]

It was up to Brigadier-General H.D.B. Ketchen's 6th Brigade to consolidate this precarious position, and improve communications through the centre of the crater sector. Ketchen's men scarcely had a chance to make good their defences when the Germans unleashed their counter-attack in the early morning hours of 6 April. They drove Ketchen's men out of their as-yet unconsolidated positions south of the craters and occupied Craters 2, 3, 4, and 5. Ketchen launched immediate counter-attacks, and later that day it was reported that elements of the 28th and 31st Battalions had retaken Craters 4 and 5. In reality, what they had done was repeat the same mistake made earlier by the British and occupied Craters 6 and 7 instead.[19] Operations continued against Craters 2 and 3, but the fact that elements of the 2nd Canadian Division unwittingly continued to occupy mistaken positions frustrated the successful attainment of their goals. Subsequent assaults on the craters were improperly coordinated and beaten back by determined enemy resistance. The German garrisons of Craters 4 and 5 were left unmolested by British and Canadian artillery fire, simply because it was believed that the two craters were in Canadian hands. Poor weather prevented effective aerial reconnaissance from 8 April onward, and it would not be until 16 April – ten days after the original error, and many casualties later – that new aerial photos would reveal the awful truth about the 2nd Canadian Division's position at St. Eloi. What had gone wrong? What had prevented the officers and men of the 2nd Canadian Division from realizing that their dispositions were incorrect? Examining the accuracy of the intelligence that was generated during the battle leads to explanations for many of the mistakes that were made.

From the beginning, the 2nd Canadian Division was handicapped by the quality of its intelligence. On 4 April, as the 6th Brigade relieved the weary British at St. Eloi, Staff Captain D. E. Macintyre confided in his diary that, "Owing to the constant fighting it had been impossible for us to send any advance parties up to reconnoitre the line. We had tried, but the English would not allow any of them in the line; [they] said they would be in the way while operations were going on. So when we took over on the night of the 3rd we took a leap in the dark." [20] Macintyre admits that they did receive maps showing where the new line was,[21] but Turner's opinion of the state of the British positions is indicative of the use that those maps must have been. In his report on operations between 3 and 7 April, he commented acidly that, "The line taken over was little more than a line on the map."[22]

Records do show that the Chief of the British General Staff, Lieutenant-General L.E. Kiggell, directed Second Army to ensure that aerial photos taken of the crater area on 31 March be shown to the relieving Canadian commanders.[23] In addition, Lieutenant-General H. Fanshawe, GOC of British V Corps, claimed in his report dated 19 April that he was satisfied with the state of the positions at St. Eloi when they were turned over to the 2nd Canadian Division on the night of 3-4 April. He reported that at the time of handing over, Colonel Smith of the 8th King's Own Lancashire Regiment, "went personally along the front line to see that the Canadians were posted" and that, "The Gordon Highlanders who were on the right of our new line, left two officers there for 24 hours after the 3rd Division had been relieved by the Canadians, and the G.S.O.1 of 3rd Division was left with the 2nd Canadian Division to assist as long as his services were required."[24] Major-General Haldane of the 3rd British Division echoed Fanshawe's opinion and maintained that, "Before handing over the command to the G.O.C. 2nd Canadian Division I satisfied myself that the whole position was occupied and in course of consolidation, and assured myself that the G.O.C. 2nd Canadian Division thoroughly understood the situation."[25]

Although it would appear that the British had done their best to inform the Canadians about their new positions, the ever-present problem of the "changing landscape" still spawned a degree of confusion. As D.E. Macintyre confessed, "The whole place was such a confused mess, that an officer that the 8th [British] Brigade left behind with us got completely lost when he undertook to show Jukes [Brigade Major, 6th Canadian Brigade] around." [26]

Despite the assistance rendered by the British during the relief, Turner was disappointed to learn that the actual positions of the opposing German infantry had not been located with any certainty.[27] Thus, while it seems that the British had done what they could to provide the 2nd Canadian Division with accurate intelligence, their alleged refusal to allow Canadian scouts into the crater area before the relief, and their inability to provide Turner and his officers with reliable information about German dispositions, severely hampered the Canadians' ability to make informed command decisions as soon as they took over the sector.

Given time, Turner's men might have been able to make good their position in the crater sector,[28] but the Germans denied them the opportunity. During 4-5 April, the Canadians were subjected to violent bombardments that destroyed much of their consolidation work before it was even finished. A British artillery officer, who had spent a year in the Ypres salient went so far as to describe the intensity of the shelling as "far greater than he had hitherto experienced."[29] The constant shelling, the pockmarked and porridge-like terrain, and the lack of discernable landmarks conspired to confound the troops as they struggled over the St. Eloi battlefield, often waist-deep in mud. To give but one example, when troops of the 2nd Canadian Pioneer Battalion were ordered to begin consolidating the craters on the night of 4-5 April, their guides lost their bearings. The men floundered about throughout the night unable to find the craters, and unable to fulfill their task.[30] Even with compasses and rudimentary maps, the men were unable to make sense of the terrain at night because each crater or shell hole looked like the next. Added to this was the physical exhaustion induced by struggling through the deep sludge. This served to disrupt the men's sense of distance regarding the amount of ground that they had traversed. Movement by daylight was virtually impossible because of the deadly accuracy and frequency of German artillery barrages, which were being directed by observers enjoying a clear view of the whole sector from atop the Wytschaete Ridge.[31] Thus, the officers and men of 2nd Canadian Division were unable to get a clear sense of the topography of the sector before the Germans launched their counter-attack on 6 April. D.E. Macintyre summed up the crux of the problem facing the Canadians in gathering accurate intelligence and in executing successful operations at St. Eloi. In his opinion:

> The 6th Brigade was suffering from the misfortune of taking part in a strictly local operation: the opening of a small gap in the line easily closed by the enemy, who could quickly stiffen his resistance at that one point and concentrate not only his own divisional artillery, but that of the flanking divisions and the heavy artillery of the Army Corps as well. The absurdity of the whole St. Eloi operation was that it only involved 600 yards of front. It was like poking a stick into a well-organized hornet's nest.[32]

A confused situation became completely chaotic in the early hours of 6 April when the Germans unleashed their counter-attack to regain the territory lost on 27 March. Although Turner lacked precise intelligence as to exactly when the

Germans would strike, the attack itself was not unexpected. Ketchen knew that trouble was brewing, and he emphasised the vulnerability of his positions on the forward slopes of the craters.[33]

In fact, the prospect of facing a German counter-attack was implicit in the reservations expressed by Turner and his superior, Lieutenant-General Alderson, regarding the plan of operations for St. Eloi as originally outlined to them by General Plumer earlier in March. Alderson and Turner had both suggested that the 2nd Canadian Division should carry out the operation from the beginning, rather than face the danger of relieving the 3rd British Division in the midst of an active battlefront. But the fact that the 3rd British Division had already trained for the operation, and the need to detonate the mines before the Germans got wind of them, compelled Plumer to veto the Canadians' suggestion. He did agree, however, that the British troops would not be relieved until they had properly consolidated their new front line.[34] The accelerated relief of the 3rd British Division before it could complete the task of consolidation therefore placed Turner in the perilous situation that he had anticipated from the beginning.

Shortly after the German guns opened up at 3:30 am on 6 April, some more precise intelligence was gained that a counter-attack was in the offing. 27th Battalion telephoned 6th Brigade headquarters to report that a prisoner had been taken. The man, believed to be a deserter from the German 214th Regiment, reported that the Germans were going to attack at dawn.[35] This message was quickly followed by the frantic call, "They are attacking now at - ..." But the conversation was cut short when the telephone line was abruptly severed.[36] This deprived both 6th Brigade and 2nd Canadian Division of their most immediate source of intelligence during the initial stages of the German counter-attack. Information now could reach headquarters only by means of runners, whose task was made extremely hazardous by the accuracy of enemy small arms and artillery fire. However, by 8:25 am, reasonably regular communication links were reestablished with forward units. Yet, continuing difficulties in maintaining consistent communications links with forward positions would diminish the intelligence gathering capabilities of the 2nd Canadian Division throughout the battle. This was an age in which easily transportable wireless sets had not yet been perfected, and the vulnerability of telephone lines would continue to bedevil commanders on both sides for the rest of the war.[37]

It was during the German counter-attack that the greatest intelligence problem arose for the 2nd Canadian Division: misidentification of the craters. After the Germans had pushed the Canadians out of their forward positions and occupied Craters 2, 3, 4, and 5, Brigadier-General Ketchen, at Turner's behest, launched a series of attacks to regain the lost ground. Attacks against Craters 2 and 3 on 6 April were unsuccessful, but, as Turner informs us in his report on 18 April, bombing attacks by elements of the 28th and 31st Battalions against Craters 4 and 5, "mistook their objective, and entered two large craters about 50 yds. short of Crater No. 5." These were Crater 6 and the older crater later identified as Crater 7. Turner goes on to say that:

> ...none of the men comprising this party had ever seen the ground before, and none had any previous experience of large craters. When they came across these craters, about 30 yards or so wide, they jumped to the conclusion that they had reached their objective, and reported accordingly. The party had suffered casualties in getting to the craters through the barrage, and they assumed that the barrage was intended to deny these craters to them.[38]

According to Turner, "This unfortunate report, which was forwarded to the Canadian Corps, was the initial cause of all the misunderstanding which has since taken place. It will be understood that the German shell fire later in the day, and on succeeding days, completely isolated the craters in question, so that no reconnoitring officers could reach them in daylight."[39] When one considers also the confounding nature of the terrain in the crater area, it is not so surprising that the attacking infantrymen were mistaken about their actual positions. Lieutenant-Colonel A.H. Bell, commanding the 31st Battalion, later wrote:

> In the accounts which have been written of St. Eloi I think that the idea which has been given is that there was one line of craters. This is by no means correct, as there were at least 17 large craters in that group. It is true that Nos. 2, 3, 4, and 5 were considerably larger than the others, but our men did not know this and any one of the 17 was much larger than

> anything of the kind that he had previously seen...Altogether the conditions were about as confusing as they well could be and I have never been in a more confusing Sector than St. Eloi.[40]

Ketchen had received conflicting reports throughout the morning of 6 April regarding the actual positions of his own men and the Germans. But when he received word at 10:00 am from the commanding officer of the 28th Battalion that his bombing party had moved through Craters 4 and 5, and that Crater 5 was now held by elements of the 31st Battalion, the situation at last appeared to be getting clearer.[41]

D.E. Macintyre was one officer who, according to his memoirs, claimed to know differently. He had been sent by Ketchen to the village of Voormezeele, behind Canadian lines, to establish an observation post and gain some accurate intelligence. At 7:20 am on 6 April, he and three brigade scouts set up shop in a ruined schoolhouse facing the craters, which were 1,400 yards away. Using a map, a compass, and two telescopes, Macintyre and his men were able to take a bearing on the craters and quickly identified German troops digging in Craters 2, 3, 4, and 5. As Macintyre explained:

> We sandbagged ourselves in near an opening in a wall, set up a map on a board, oriented it with the country by the aid of a compass, took a bearing on the craters, aligned two telescopes on them and went to work.
>
> We soon had the old Boche located, digging like good fellows on the sky-line of the craters. They were exposed from the waist up and nobody seemed to be annoying them. I saw them digging in craters 2, 3, 4, and 5, and went to 28th Battalion headquarters to report what I had seen. General Ketchen, however, was receiving reports from other sources telling him that our men were still in and out in front of craters 2 and 3 at any rate, so he was loath to order that they be shelled. No doubt some of the reports reaching him from the actual front were so relayed, and delayed, that by the time they reached him the news was cold...Of course, I could not see what our troops were doing, but I certainly could see the Germans industriously digging in, and could not get anybody to believe me.

> Scouts Parker, Griffin and Lindsay were with me and they saw what I saw. They were all trained observers... They, like myself recognized the German uniforms and even the shovels they used. In fact, we could see their faces, and I have a distinct recollection of one fellow pausing in his labour and looking our way with a grin.
>
> All day long we watched them dig, till they were dug in up to their necks. Not until later in the war could we get Canadians to dig like that.[42]

Brigadier-General Ketchen claimed that Macintyre "reported early on that morning that the enemy were apparently in Craters 4 and 5. This was passed on to the Division. Later in the morning Capt. Macintyre reported that no movement could be seen in these two craters at all, but that the Bosche was working hard in Craters 2 and 3."[43] Indeed, Macintyre's report on the events of 6 April, which he submitted to divisional headquarters on 9 April, gives the impression that he believed Craters 4 and 5 were not occupied by the Germans later in the day.[44] However, in his postwar account of the battle, Macintyre claims that he "had maintained from the first that the Germans had entered and consolidated [Craters] 4 and 5 on the 6th..."[45] This assertion possibly represents the conviction of hindsight, rather than the facts as he reported at the time.

Although we may never know the exact truth, a reasonable explanation for the contradictions between reports from the front line and Macintyre's observation post farther to the rear is provided by Lieutenant-Colonel A.H. Bell, who, after the battle, spent some time "studying the crater area from different angles and distances." One of the vantage points taken up by Bell during this personal investigation was similar to that occupied by Macintyre on the morning of 6 April. Bell noted that while "observing with field glasses from the high ground behind Voormezeele, one could clearly trace the main line of craters." However, the perspective changed when one examined the same ground from a lower and more forward position:

> From the front line, as it was held after the battle, you saw what appeared to be the same line; but you were really looking up at the smaller craters in front and obscuring the view of the main line [Craters 2, 3, 4, and 5]. Officers and men gained quite erroneous impressions regarding their positions at certain times and passed inaccurate accounts to correspondents in perfect good faith.[46]

The error made by troops of the 6th Brigade on 6 April was repeated in subsequent days by men of the 4th and 5th Brigades as they did their tours in the front line. Not even the reports of three German prisoners, who had been captured in the early morning hours of 7 April, and who claimed unanimously that the Germans held Craters 2, 3, 4, and 5 with about 70 men in each, were able to convince the Canadian commanders of the truth.[47] This was because Turner and his brigadiers continued to receive regular reports from their own front lines that Craters 4 and 5 were secured and being used, along with Crater 1, as bases of operations against enemy-held Craters 2 and 3.

In fact, from Turner's perspective, the situation must have seemed progressively rosier, because starting on 9 April, reports began to trickle in that troops of Brigadier-General Robert Rennie's 4th Brigade had made lodgments in both Craters 2 and 3.[48] At last it appeared that the situation might soon be well in hand. The encouraging news was communicated to Lieutenant-General Alderson at Canadian Corps headquarters, and Alderson made sure that these reports were verified before submitting them to General Plumer. Unfortunately, by 8:00 am on 11 April, fresh reports arrived that cast serious doubt on the optimistic news of the previous two days. That same day, Alderson sheepishly reported to Plumer:

> There is no doubt that the situation at present is that our line runs from the original line through Crater 6 to Crater 5, encircles Crater 5, thence to Crater 4, encircles Crater 4, with a good 5-ft. deep trench dug last night running just North of Craters 4 and 5.
>
> None of our troops are in Craters 2 and 3, but about 40 of the 18th Battalion are in a trench 75 yards north of Crater 3…It is not known whether the Germans hold Craters 2 and 3, but it is certain that they are occupying the trench just south of the Craters.
>
> The barrage which the enemy puts between the Craters and St. Eloi is intense and had caused much confusion and loss each night. The whole place is a mass of craters and enormous shell holes full of water and mud, and movement is very difficult, and it is certain from the C.O.'s accounts that troops after falling from one hole into another do not know where they are. The holes are very deep and the men looking up against the sky line think they have reached the craters when they have not…

> I am confident that Br. General Rennie has done all in his power to gain accurate information and that the reports from the front, with which communication at night is only possible, have misled him[49]

In his reply, Plumer expressed his regret, "that reports which it is now evident were inaccurate should have been transmitted,"[50] and the disappointing news was sent further up the chain of command to the Commander-in-Chief. This triggered a stinging rebuke from Field Marshal Sir Douglas Haig, who informed all concerned of "the necessity for superior commanders, especially in the case of the comparatively new units at present in our Armies, to take steps at once to satisfy themselves that all that is necessary, advisable and possible has been done. The principle of leaving execution entirely to subordinates may easily be carried too far in dealing with troops and subordinate officers who are far from being highly trained or experienced."[51]

In order to confirm once and for all the location of the Canadian positions, Alderson dispatched one of his own intelligence officers, Lieutenant S.A. Vernon, to make a personal reconnaissance of the St. Eloi battleground. Vernon conducted his inspection on the night of 11-12 April, and he covered what he thought was the entire front line. In his report he confirmed that the Germans positively held Craters 2 and 3, and that the Canadians held Craters 4 and 5.[52] Any doubts about the Canadians' positions were now seemingly dispelled. Considering the intelligence that he was getting from his own brigades, as well as from Canadian Corps headquarters, it becomes difficult to fault Turner for basing his decisions upon the incorrect premise that his men held all but Craters 2 and 3.

There is one final but important matter to address in assessing the effect of intelligence upon operations at St. Eloi. That is the question of aerial intelligence. It was aerial photographs that were taken on 16 April that finally cleared up the situation and revealed the awful truth that 2nd Canadian Division had been occupying the wrong craters since 6 April. Why did it take so long for aerial intelligence to disclose the truth? The reason lay in poor weather conditions that plagued the region throughout operations. Periods of rain, wind, fog, and low cloud effectively hampered air operations during much of the battle. But the weather cleared up enough on 8 April to allow a plane to get up and photograph the crater sector. The great mystery to most historians is why the aerial photo taken on 8 April did not reveal to Turner and his staff the reality of their positions after 6 April.[53] The photo shows Craters 6 and 7 (the craters actually occupied by the Canadians, rather than Craters 4 and 5) to be partially filled with water, while the other craters appear comparatively dry.[54] One would think that this should have been an obvious means of identification, but D.E. Macintyre offers a possible solution as to how this photo might have been misinterpreted. As he claims in his postwar memoirs, the photo taken on 8 April:

> ...showed a newly dug trench around craters 4 and 5, and this was mistaken at first for the trenches which our troops were supposed to have dug. The ineffectual ditches which had been attempted around 6 and 7 were overlooked. We congratulated ourselves on the splendid work that we thought had been done by our crater garrisons. It was conceded finally that the enemy were in 2 and 3, but it was thought by late afternoon that we had occupied 4 and 5. I had maintained from the first that the Germans had entered and consolidated 4 and 5 on the 6th, but it was not until April 16th that we got really clear photographs that showed a continuous trench system had been dug by the enemy connecting up all of the four large craters, and that we had done some work in trying to connect up 6 and 7. Of course, it is easy to understand now how a soldier looking out over the rim of 6 could see 5 and mistakenly think he was in 4 and observing 3. A better understanding of airplane photography, which showed crater 7 to be half full of water while the others were comparatively dry, would have saved a lot of trouble, but at that time few of our officers had ever seen air photos and knew little about the reading of them.[55]

Macintyre's additional assertion that the quality of the 8 April photo was "rather poor" [56] also may have contributed to the misinterpretation.

Of course, accepting this proposition requires taking Macintyre at his word. Yet, his explanation is believable for several reasons. For one thing, aerial photography and photo interpretation were, in early 1916, evolving sciences. Although various notes and instructions on how to read aerial photos had been in circulation since at least the fall of 1915,[57] the first comprehensive manual for interpreting air photos would not be published until January 1917.[58] Furthermore, it was during the battle for the St. Eloi craters that the Royal Flying Corps (RFC) began reorganizing its system for processing and distributing aerial photographs.

By the spring of 1916, demands for air photos were becoming so great that the corps wing photographic sections were becoming overburdened with work, and this resulted in delays in the delivery of prints. As the Royal Air Force Official History explains, "There were occasions when prints reached the units too late for full advantage to be taken of the new information they revealed."[59] In order to speed up the delivery of prints, the RFC established small photographic sections in each corps squadron and in each army reconnaissance squadron. This decentralization process began in the middle of April 1916. With this new arrangement, the staff of each army corps submitted orders directly to the commander of the RFC squadron attached to that corps for all photographs of enemy positions taken along the corps front.[60] Since the records are clear that the photo taken on 8 April *did* reach the units concerned in a timely fashion, there is no reason to assume that any delays in the delivery of air photos affected operations at St. Eloi.

What is of interest, however, is the fact that divisions did not order photos directly from RFC squadrons. Before the reorganization described above, which took place after the battle for the craters was effectively over, photos would have come from the corps wing photographic section, which was even farther removed from the division than were the individual squadrons that were attached to each army corps. In any case, photos would have been supplied to divisions by the corps staff. This means that not only were Turner and his staff responsible for misinterpreting the 8 April photo, but also personnel at Canadian Corps headquarters bore some blame as well. Support for this notion can be gotten from no less than Field Marshal Sir Douglas Haig. In the 17 April entry of his diary Haig states:

> After lunch the CGS [Chief of the General Staff] reported to me. He told me that General Plumer had come to GHQ [General Headquarters] in such low spirits, that he would not come to lunch or see me unless I asked for him. The reason was that an air photograph taken yesterday had disclosed the fact that the Canadians were not holding the craters at

> St. Eloi, and it seemed doubtful if they ever held them, because the new photo showed each crater strongly fortified by the enemy and a continuous line in front joining them all up together. In some parts the Canadians are back in over our original line! It was not until the photographs reached Corps HQ that the situation was realized and inquiries instituted...[61]

In his letter to General Plumer on 16 April, Lieutenant-General Alderson described how the aerial photo taken earlier that day cleared up the confusion, and he admitted his own role in verifying the photo's implications. He also presented a succinct summary of the reasons for the muddle, which accorded well with those provided by Turner in his own report on 18 April. Regarding the erroneous dispositions of Turner's men, Alderson wrote:

> It is very distressing to the Divisional Commander and myself to find that this is the state of affairs but I have had the photograph verified by the R.E. [Royal Engineer] Officer who has been in charge of the work and it leaves no doubt on the point. His work is all shown on the photograph. The mistake arose through the original report sent in when the forward line was lost in which it was stated that we still held Crater 5.
>
> This was in reality Crater 6. Troops were new to the ground and no troops had been in the Craters, and there was nothing to distinguish one Crater from another. Starting from the assumption that Crater 6 was Crater 5, several Staff Officers, Regimental Officers and R.E. Officers have carried out reconnaissances from that point and all work has been carried out from this basis...
>
> I very much regret this unfortunate mistake, more especially as it has quite unintentionally misled you as well as myself throughout the whole of the operations.[62]

Therefore, a case can be made that Canadian Corps staff also bear some measure of responsibility for the interpretation of aerial intelligence. It is possible that if they were among the first to analyze the photos before passing them along to 2nd Canadian Division, they might also have passed along their own interpretations of those images with the photos themselves. This may have influenced 2nd Division's interpretation of the photos and helped perpetuate the misconceptions surrounding Canadian dispositions in the crater sector.

Having now examined the nature of the intelligence that was reaching Turner and his staff during operations at St. Eloi, we are in a better position to appreciate the level of his responsibility for failing to recognize sooner that his men were occupying the wrong positions. In their studies of the course of operations at St. Eloi and their aftermath, historians Thomas Leppard and Tim Cook sharply criticize Turner for failing to maintain an effective grip on the situation and for misreading the information that reached him. Cook believes it is, "too easy to exonerate Turner and his Brigadiers for the poor handling of the battle by blaming the inaccurate information they received from the front…[and that] a more diligent attempt should have been made to understand the situation."[63]

However, an examination of 2nd Division's War Diary shows that virtually all reports from the front lines confirmed the opinion that the Canadians held Craters 4 and 5. In terms of the evidence reported in the divisional War Diary, from 7 April onward, any uncertainties regarding the craters reflect the status of Craters 2 and 3, and *not* Craters 4 and 5. Aside from the aerial photo of 8 April, research thus far has only been able to turn up two reports between 7-15 April that may have cast doubt on the notion that Turner's men occupied Craters 4 and 5, and both are found in G.W.L. Nicholson's Official History. Nicholson claims that, "Although as early as 10 April one of Turner's staff officers reported having been fired on from Crater No. 4, the belief that only Craters 2 and 3 were in German hands persisted."[64] Unfortunately, Nicholson does not provide his source for this information, and a search of the War Diary and accompanying Daily Intelligence Summaries has so far failed to turn up any reference to this occurrence. The second report cited by Nicholson concerns a reconnaissance made on the night of 14-15 April by Major J.A. Ross and Lieutenant Greenshields of the 24th Battalion. Nicholson maintains that they made "a personal reconnaissance of Craters 2, 3, 4 and 5 and found all four in the possession of the enemy. This was confirmed on the 16th by aerial photography…"[65] The first reference to this report in 2nd Division's War Diary is found in the Daily Intelligence Summary for 16 April. But this version of the report contains no reference to the situation as described by Nicholson. It refers only to reconnaissance work around Crater 3.[66] So, unless intelligence to the contrary was omitted from the divisional War Diary and Daily Intelligence Summaries, one can only assume that as far as Turner knew, his men were in possession of both Craters 4 and 5 because that is what he was repeatedly told. A scan of references in the divisional War Diary and Daily Intelligence Summaries turns up over twenty reports between 7 and 15 April that indicate Canadian possession of Craters 4 and 5, with no clear reports to the contrary. Turner simply was not given enough intelligence to warrant seriously questioning the fact that his men held Craters 4 and 5 – an assumption that would have been supported by Lieutenant Vernon's personal reconnaissance on the night of 11-12 April. At St. Eloi, the intelligence that Turner received regarding the dispositions of his own men was of such uniformly faulty character that it is little wonder his plans to retake Craters 2 and 3 met with failure.

The matter of the misinterpreted aerial photo of 8 April is much more difficult to explain. D.E. Macintyre's statement to the effect that many Canadian officers in early 1916 were not skilled in reading photos could possibly account for this. On the other hand, historian Tim Cook suggests that Turner and his staff may not have examined the photo as rigorously as they should have, "because they thought there was nothing to search for."[67] They most likely assumed that the trenches shown around Craters 4 and 5 represented their own consolidation efforts and not those of the Germans. And, as stated earlier, Macintyre's contention that the 8 April photo was not so clear as those taken on 16 April also may have affected their judgement. The misreading of the 8 April aerial photo stands among the most serious intelligence blunders on the part of Turner and his staff, but they can certainly share the blame for this mistake with their superiors at Canadian Corps headquarters.

When the aerial photos of 16 April revealed the fact that Turner's men did not occupy Craters 4 and 5, further operations were suspended and the search for a scapegoat upon which to pin the blunder began. Alderson eventually recommended that both Turner and Ketchen be removed from their commands, but Canadian military and political exigencies made it difficult for the British High Command to fire any Canadian general officers. Eventually, it was found to be more politically expedient to retain both Turner and Ketchen, and instead remove Alderson.[68]

Turner and Alderson had a strained relationship ever since the Second Battle of Ypres in 1915, and it would have been impossible for them to continue working together after Alderson had submitted a letter to Plumer on 18 April in which he called for Turner's removal. Oddly enough, in this letter Alderson offers no specific examples of how Turner might have blundered at St. Eloi. Rather dryly, Alderson explains, that he did not "consider that the operations reported on are, in themselves, a sufficient reason [for Turner's dismissal] (as that would be condemning a man for one mistake, which although responsible for, he had not directly committed himself)..." Instead, Alderson invoked an earlier negative appraisal of Turner's fitness for command that he had expressed in a letter to GHQ on 23 June 1915, which "was to the general effect that I did not consider that he had the necessary qualifications for a Divisional Commander, beyond the fact that he was, as testified by his Victoria Cross, physically brave to a fault."[69]

Ironically it was Alderson who was forced to relinquish his command, in part because of the political machinations of the Canadian Minister of Militia, Sir Sam Hughes, and the General Representative for Canada at the Front, Sir Max Aitken. These men helped convince Sir Douglas Haig that the cooperative relations between Canada and Britain would be soured should Turner and Ketchen be dismissed. On 28 May 1916, Alderson was "kicked upstairs" to become Inspector General of Canadian Forces in England, and was replaced by Sir Julian Byng as commander of the Canadian Corps.[70] Turner, on the other hand, retained his position as GOC, 2nd Canadian Division until December 1916, when he was promoted to command the Canadian Forces in the British Isles.

Although Turner ultimately emerged unbroken from the affair, the catastrophe at St. Eloi would continue to haunt his reputation for years to come. After the war, and with a good deal of bitterness, Turner wrote to Colonel A.F. Duguid, complaining:

> I would like to bring out this was a poorly conceived operation; in view of the Germans' dominant position at Wytschaete…
>
> We lost a large number of valuable lives over this fool business – which represented poor staff work on the part of the 2nd Army *and* poor support from Alderson when my objections were made to him. The mean

> attempt to foist the blame for lack of permanent success on my Battalion C.O.'s, Brigadiers – and finally myself was unforgivable.[71]

In Plumer's defence, Sir C.H. Harington (GSO1, Canadian Corps during the battle for the craters) remarked to Duguid, "I think Gen. Turner's comments are in some cases unfair. As I know from long experience later, Gen. Plumer did not commit troops without knowledge. He knew the St. Eloi Sector during the early fighting of 27th and 28th Divisions long before Gen. Turner came to France..."[72] As for the responsibility for passing on incorrect intelligence regarding possession of Craters 4 and 5, Harington contended, "We very rightly incurred the wrath of the Army and G.H.Q. for giving wrong information. The responsibility rested with the Canadians of which I had the honour to be B.G.G.S. [Brigadier-General, General Staff]".[73]

Colonel A.F. Duguid, in his notes for the unpublished second volume of his Official History left not one, but two personal appraisals of the debacle at St. Eloi. Regrettably, neither of these appraisals is dated, so it is impossible to tell for certain which one was written first. One of them is handwritten while the other is typed. As such, it is possible that the typed appraisal is later in date. In the final paragraphs of the handwritten appraisal, Duguid concludes:

> Seeing that, on or after the 6th, the enemy had never been driven out of any of the craters, it should have been obvious that they must still be there: a fact which the Army map, – prepared from air photos taken on the 8th and issued on the 9th – makes abundantly clear.
>
> Reluctance on the part of the 2nd Canadian Division – individually and collectively – to admit even to themselves that the Craters entrusted to their charge had been lost must surely have coloured the vision and distorted the judgement of the commanders and staffs who, looking at the photograph of the 8th, and the map based on it, and the terrain itself, so grossly misread the evidence. Such an attitude of mind evidently impelled them to disbelieve, discount or discard everything pessimistic, and to seize upon anything favourable in reports from the front line.[74]

Given the weight of evidence from divisional records and officers' reports, Duguid's statement that, "on or after the 6th, the enemy had never been driven out of any of the craters," smacks completely of judgement through hindsight. Virtually all front line intelligence from late morning on 6 April onwards, maintained that the Germans only held Craters 2 and 3. It was by no means obvious to Turner, his subordinate commanders, or to his superiors that the Germans still held Craters 4 and 5 as well. D.E. Macintyre's explanation for the misreading of the 8 April photo offers a possible reason why the situation, as laid out on the Army map of 9 April, was not "abundantly clear." The fact that Turner, his subordinates, and superiors were on 11 April able to admit their mistake in believing earlier erroneous front line reports regarding the status of Craters 2 and 3, shows that they were willing to believe negative reports when confronted with enough evidence. The volume of reports denying Canadian possession of Craters 4 and 5 was too miniscule to be considered sufficient evidence at the time. If Turner still had any lingering doubts about the security of his putative positions in Craters 1, 4, 5, and 6, they most likely were dispelled by Lieutenant Vernon's reconnaissance on the night of 11-12 April. The fact that Craters 4 and 5 remained in enemy hands became clear only after new aerial photos were taken on 16 April.

Duguid's typed appraisal, on the other hand, is much less damning in its verdict. In this, presumably later version, the Official Historian maintained that:

> From the outset the project was ill conceived and must surely have been ill starred: the sector chosen gave the enemy every advantage of observation and position, allowing him to develop to the full the use of the arm which, unless neutralized in such a fight, must rule the battle – heavy artillery.
>
> In the estimation of the enemy the Canadian infantryman was as a fighter unexcelled, and his physical condition should have been better, for his ration was better. At this stage, however, he was not as effective in trench warfare as the German for he was exhausted before the actual fight began. The British had not applied their engineering skill to the amelioration of natural conditions to the same extent, nor was the art of constructing habitable trenches so well understood or so assiduously practised by the pioneers and those others who had to occupy them.
>
> The machine guns and trench mortars, handicapped by inconsistent systems of field defence, bad communications and wet trenches, and

> busy learning in the face of the enemy to use new weapons, while also becoming accustomed to changed organization and control did not approach their full fighting value. The British field artillery was generally superior to the German, though here at a disadvantage as to observation and choice of targets. The British heavy artillery was outclassed by the German in number of guns, weight and quantity of ammunition, observation and application of fire.
>
> The apportionment of an exact relative value to each of these factors is arbitrary; but in the final analysis, all excuses are met by the fact that two German battalions well covered by artillery, retook and held the craters.[75]

This is a much kinder appraisal in terms of the account that it takes of weather, terrain, communications, inexperience, and German advantages in observation and artillery. Duguid is not as kind to the British as he could be when says that they "had not applied their engineering skill" in effectively consolidating their original gains. As we have seen, the 3rd British Division was depleted, exhausted, and subjected to severe artillery bombardments that destroyed what work they did manage to carry out. They were not even in good enough condition to carry out their dead and wounded when they were relieved on the night of 3-4 April. If fresh troops of the 2nd Canadian Division were unable to consolidate their positions and hold off the German counter-attack on 6 April, how much worse would the men of the 3rd British Division have fared if they had been left in the line until they were originally scheduled to be relieved on the night of 6-7 April?

Some of the mysteries surrounding Canadian operations at St. Eloi may never be explained conclusively. However, a critical examination of the recorded intelligence that reached Turner, his staff, subordinates, and his superiors, leads one to conclude that assigning individual blame for the fiasco is a futile endeavour. Of course, Turner, as GOC of the 2nd Canadian Division, must be held accountable for the actions of all those who served under his command. But the chain of responsibility should not have to end with Turner, for Turner's decisions were predicated as much upon the orders and intelligence supplied by his superiors as they did upon the information provided by his subordinates.

As the planners of an operation that was arguably flawed in its initial conception, Plumer and his Second Army staff have to bear a considerable amount of responsibility. Even if the Canadians had made no errors in ascertaining their own dispositions, the narrow front of the original British attack and the clear advantages enjoyed by German artillery observers made the position in the St. Eloi crater sector terribly vulnerable, and perhaps even untenable.

In addition, Alderson and the intelligence staff at Canadian Corps can be held at least partially accountable for the errors in interpreting aerial intelligence, as well as for the reconnaissance report of Lieutenant Vernon on the night of 11-12 April. In fact, in his report to Plumer on 18 April, Alderson allowed that Turner and his officers had done their best, and even conceded a certain share of personal responsibility. Alderson admitted that he knew, "that both Divisional and Brigade Staff Officers have tried to clear up the situation and I have sent officers of my own staff down for the same purpose and to see that the work of restoring our defences was being pushed on. I and my B.G.G.S. have visited the Division and Advanced Brigade H.Q. daily and satisfied myself that the situation was exactly as represented."[76]

But despite the fact that Alderson and Turner had been misled equally, the Corps Commander maintained that final responsibility for the deadly confusion had to rest with Turner and the 2nd Canadian Division. Alderson pronounced that, "In spite of the difficulties...I am of the opinion that it should not have taken 10 days for the Division concerned to have ascertained the exact situation of its troops, and the responsibility for the forwarding of inaccurate information during this period must rest with the 2nd Canadian Division."[77]

Even though Turner was the commander most directly responsible for the operations at St. Eloi, the preceding analysis clearly demonstrates that his conduct of the battle was based upon consistently faulty intelligence which resulted from fighting conditions of the harshest kind, from inexperienced troops, and also from an intelligence organization that was still growing, evolving, and learning its trade. The establishments and duties of intelligence personnel in the Canadian Corps, as laid down in 1918, were not so well articulated in early 1916.[78] Thus, mistakes were more likely to occur during the earlier years of the war. When coupled with the variety of obstacles facing him (weather, terrain, equipment problems, German artillery), the odds became so heavily stacked against Turner that it is hard to imagine how he could have performed any better. Although Turner may not have been among the most effective Canadian division commanders during the First World War, when we consider the formidable nature of operations at St. Eloi, combined with the provision of inaccurate intelligence, it is difficult to say whether or not even the most talented of commanders would have fared any better. If "the buck" must stop with Turner for his division's failure at St. Eloi, it stops with him only because it was primarily a divisional action for which he happened to be the General Officer Commanding.

During the battle for the St. Eloi craters the contradictions in intelligence did not balance each other out in Clausewitzian fashion. Unfortunately, they were weighted too heavily on the inaccurate side. This led to misperception on the part of commanders at all levels. Nevertheless, the politics of command demanded payment for failure. Alderson paid the price with his command and Turner with another blow to his already questionable reputation.

---

1 Carl von Clausewitz, *On War*, trans. Michael Howard and Peter Paret (Princeton: Princeton University Press, 1984), p.117.

2 Colonel G.W.L. Nicholson, *Canadian Expeditionary Force, 1914-1919* (Ottawa: Queen's Printer, 1962), p.144.

3 Among the Canadian officers that were "degummed" following operations at St. Eloi were: Lieutenant-Colonel E.W.S. Wigle, 18th Bn.; Lieutenant-Colonel J.A. MacLaren, 19th Bn.; Lieutenant-Colonel J.L. McAvity, 26th Bn.; Lieutenant-Colonel I.R. Snider, 27th Bn. These four battalion commanders were, in Turner's opinion, too old for command. Lieutenant-Colonel C.A. Ker, G.S.O. (1) of 2nd Canadian Division, and a British officer, also left. He was transferred to a British division on May 24, 1916.

4 Until the 1990s, the principal published sources for the battle at St. Eloi were: Lord Beaverbrook, *Canada in Flanders, Volume II* (London: Hodder and Stoughton, 1917); James E. Edmonds, *History of the Great War: Military Operations, France and Belgium, 1916* (London: Macmillan, 1932); and G.W.L. Nicholson, *Canadian Expeditionary Force, 1914-1919* (Ottawa: Queen's Printer, 1962).

5 See Thomas P. Leppard, "Richard Turner and the Battle of St. Eloi," M.A. Thesis, University of Calgary, 1994; and Tim Cook, "The Blind Leading the Blind: The Battle of the St. Eloi

Craters," *Canadian Military History* 5, 2 (1996). See also Leppard's article, " 'The Dashing Subaltern' Sir Richard Turner in Retrospect," *Canadian Military History* 6, 2 (1997).

6 Edmonds, British Official History, pp.178-179.

7 According to Canadian Official Historian Colonel A.F. Duguid, the activities of opposing British and German miners in the St. Eloi area resulted in the blowing of thirty-three mines and thirty-one camouflets in an area of ten acres. Library and Archives Canada (LAC) Record Group (RG) 24, C-6-j, Vol. 6992, Vol. II, Chapter X, para. 3 of the unpublished Canadian Official History, Vol. II.

8 In a preliminary memorandum issued by the 9th British Infantry Brigade on 18 March 1916, it is stated quite clearly that, "The object of the forthcoming operation is to cut the Enemy's Salient and thereby gain important ground for observation and also to destroy the Enemy's extensive Mining system which is known to exist." LAC, RG 9, III-D-3, Vol. 5070, File titled, "3rd Division; Papers relating to operations at St. Eloi, 27th March to 3rd April 1916."

9 LAC, RG 24, C-6-j, Vol. 6992, Vol. II, Chapter X, para. 16 of the unpublished Canadian Official History, Vol. II.

10 Edmonds, British Official History, p.183.

11 LAC, RG 9, III-D-3, Vol. 5073, File No. 1, "Summary of Operations – Second Army for the period 22nd-31st March, 1916," p.4; and LAC, RG 9, III-D-3, Vol. 5074, File No. 5, "Summary of Reports on the Operations at St. Eloi – March 27th to April 18th, 1916," pp.3-6.

12 Edmonds, British Official History, pp.184-185.

13 LAC, RG 9, III-D-3, Vol. 4843, War Diary, 2nd Canadian Division Headquarters, March 1916, 2nd Canadian Division Order No. 37, 23 March 1916; ibid., April 1916, Appendix 177, "Narrative of Events up to the Occupation of the St. Eloi Position," 12 April 1916.

14 LAC, RG 24, C-6-j, Vol. 6992, Vol. II, Chapter X, para. 20 of the unpublished Canadian Official History, Vol. II.

15 Nicholson, p.140.

16 Beaverbrook, p.126.

17 LAC, MG 30, E 241, D.E. Macintyre Papers, Vol. 2, manuscript for *Men of Valour*, p.37.

18 Nicholson, p.140.

19 LAC, RG 9, III, D3, Vol. 4843, R.E.W. Turner, "Report on 2nd Canadian Division's Operations at St. Eloi. April 3rd to April 16th," 18 April 1916, pp.5-6.

20 LAC, MG 30, E 241 D.E. Macintyre Papers, Vol. I, Personal diary, 4 April 1916, p. 142. An intriguing remark by Brigadier-General Ketchen would appear to contradict Macintyre's assertion. In an interview on 9 April 1916, Ketchen stated that he returned from England on 1 April and received word that his Brigade was to take over the St. Eloi salient. He claimed that, "The ground was examined prior to taking over. Officers of the 27th Bn. had reconnoitred the ground on the left, officers of the 31st the trenches on the right. Later, the 2nd Division had reversed this arrangement, so that when eventually the Battalions moved in, they were unfamiliar

with the ground in front of them." LAC, RG 9, III-D-1, Vol. 4679, Folder 14, File 1, Notes by Wm F. Bradley on the action at St. Eloi.

21 LAC, MG 30, E 241 D.E. Macintyre Papers, Vol. I, Personal diary, 4 April 1916, p.142.

22 LAC, RG 9, III-D-3, Vol. 4843, R.E.W. Turner, "Report on Operations of the 2nd Canadian Division After Arrival in the New Divisional Area St. Eloi. From April 3rd to April 7th, 1916," p.1.

23 LAC, RG 9, III-D-3, Vol. 5074, File No. 6, G.H.Q. O.A.D. 677, to Second Army, 3 April 1916.

24 Ibid., V Corps G.X. 5303/22 to Second Army, 19 April 1916.

25 Ibid., File No. 5, 3rd Division, G. 1230 to V Corps, 17 April 1916.

26 LAC, MG 30, E 241, D.E. Macintyre Papers, Vol. I, Personal diary, 4 April 1916, p.142.

27 LAC, RG 9, III-D-3, Vol. 4843, R.E.W. Turner, "Report on 2nd Canadian Division's Operations at St. Eloi. April 3rd to April 16th," 18 April 1916, p.2.

28 Turner believed that, "if we were given time, we could make a pretty good line of the position selected, but that a very great amount of work would be required; also that the captured line was such a short one that it formed an ideal objective for the concentrated artillery fire of the Germans." LAC, RG 9, III-D-3, Vol. 4843, R.E.W. Turner, "Report on 2nd Canadian Division's Operations at St. Eloi. April 3rd to April 16th," 18 April 1916, p.3.

29 Ibid.

30 LAC, MG 30, E 241, D.E. Macintyre Papers; Vol. I, Personal diary, 4 April 1916, p.143.

31 D.E Macintyre attempted to get around the entire front on the morning of 4 April, but the German artillery put a stop to this endeavour. Ibid., Vol. II, manuscript for *Men of Valour*, pp.35-36. A similar attempt, made on 5 April by Brigadier-General Ketchen and several staff officers, met with equal success for the same reason, and almost resulted in Ketchen being killed. LAC, RG 9, III-D-3, Vol. 4843, R.E.W. Turner, "Report on 2nd Canadian Division's Operations at St. Eloi. April 3rd to April 16th," 18 April 1916, p.3.

32 LAC, MG 30, E 241, D.E. Macintyre Papers, Vol. II, manuscript for *Men of Valour*, p.39.

33 LAC, RG 24, C-6-a, Vol. 1739, File DHS 3-17 (vol. 4), H.D.B. Ketchen, "The Actions of St. Eloi Craters. 27th March-16th April, 1916," 14 February 1929.

34 Nicholson, p.138. In a handwritten personal appraisal composed after the battle, Turner maintained, "When I heard we had to take over St. Eloi, I asked to have operations delayed, and allow us to take them on, as I considered it sounder. It is one thing to take a position, and another to hold it. I did not originate the situation which developed and has cost 1600 men." At the top of this document, in what appears to be Colonel A.F. Duguid's handwriting, is the heading, "Written by Gen. Turner. Probably 18th Ap. 1916." LAC, RG 9 III-D-1, Vol. 4679, Folder 14, File 1.

35 LAC, RG 9, III-D-3, Vol. 4812, Canadian Corps General Staff, War Diary, "Canadian Corps Summary of Intelligence," 6 April 1916, pp.2, 4.

36 LAC, MG 30, E 241, D.E. Macintyre Papers, Vol. II, manuscript for *Men of Valour*, p.40.

37 LAC, RG 9, III-D-3, Vol. 4843, page 3 of H.D.B. Ketchen's report on operations of 6 April, dated 12 April 1916, and labelled "Appendix No. 1" in R.E.W. Turner's, "Report on 2nd Canadian Division's Operations at St. Eloi. April 3rd to April 16th." 18 April 1916.

38 LAC, RG 9, III-D-3, Vol. 4843, R.E.W. Turner, "Report on 2nd Canadian Division's Operations at St. Eloi. April 3rd to April 16th." 18 April 1916, pp.5-6. D.E. Macintyre offers an explanation for this lack of familiarity with the terrain for attackers from the 28th Battalion. According to Macintyre, "The 28th were now the only unit in the brigade that had not been used in the front line, and as they had two companies in support in Voormezeele close by, Lt.-Col. Embury was ordered to send in a strong party of bombers who were to make a determined effort to take and hold Craters 5 and 4." LAC, MG 30, E 241, D.E. Macintyre Papers, Vol. II, manuscript for *Men of Valour*, p.45.

39 LAC, RG 9, III-D-3, Vol. 4843, R.E.W. Turner, "Report on 2nd Canadian Division's Operations at St. Eloi. April 3rd to April 16th." 18 April 1916, p.6.

40 Public Record Office (hereafter, PRO) Cab 45/149, Postwar British *Official History* Correspondence and Accounts, St. Eloi, Personal Account of Brigadier-General A.H. Bell, pp.6, 7.

41 LAC, RG 9, III-D-3, Vol. 4843, page 3 of H.D.B. Ketchen's report on operations of 6 April, dated 12 April 1916, and labelled "Appendix No. 1" in R.E.W. Turner's, "Report on 2nd Canadian Division's Operations at St. Eloi. April 3rd to April 16th." 18 April 1916.

42 LAC, RG 30, E 241, D.E. Macintyre Papers, Vol. 2, unpublished manuscript, *Men of Valour*, pp.41-42. See also "Report from Captain D.E. McIntyre [sic], Staff Captain, 6th Canadian Inf. Bde. To G.O.C. 2nd Canadian Division," 9 April 1916, in LAC, MG 30, E 46, Sir Richard Turner Papers, Vol. 2, File 10. This report is perplexing because after the initial admission that Macintyre observed Germans in Craters 2, 3, 4 and 5, for the rest of the report his specific references to German-held craters name only Craters 2 and 3. He does not explicitly confirm or deny that he still believed that the Germans continued to hold Craters 4 and 5.

43 LAC, RG 9, III-D-1, Vol. 4688, Folder 42, File 15, Ketchen to Sir Max Aitken, 28 July 1916.

44 LAC, MG 30, E 46, Sir Richard Turner Papers, Vol. 2, File 10, "Report from Captain D.E. McIntyre [sic], Staff Captain, 6th Canadian Inf. Bde. To G.O.C. 2nd Canadian Division," 9 April 1916.

45 LAC, MG 30, E 241, D.E. Macintyre Papers, Vol. 2, unpublished manuscript for *Men of Valour*, p.47.

46 PRO, Cab 45/149, Postwar British *Official History* Correspondence and Accounts, St. Eloi, Personal Account of Brigadier-General A.H. Bell, p. 11.

47 LAC, RG 9, III-D-3, Vol. 4812, Canadian Corps General Staff, War Diary, "Summary of Operations, Canadian Corps, 1st To 7th April 1916 inclusive." In the Canadian Corps Intelligence Summary for 7 April, it is stated that, "All three prisoners claim that the Germans occupied the craters Nos. 2, 3, 4 and 5 up to midnight April 6/7th. (at present moment, noon April 7th, this is not the case…)." LAC, RG 9, III-D-3, Vol. 4812, Canadian Corps General Staff, War Diary,

"Examination of Prisoners Captured at St. Eloi Craters on the Morning of 7th April 1916," Appendix to "Canadian Corps Summary of Intelligence," 7 April 1916.

48 LAC, RG 9, III-D-3, Vol. 4843, 2nd Canadian Division General Staff War Diary and Daily Intelligence Summaries for 9, 10 April 1916.

49 LAC, RG 24, C-6-j, Vol. 6992, Vol. II, Chapter XI, Copy of Canadian Corps G. 120 to Second Army, April 11, 1916.

50 Ibid., Memo to Commander Canadian Corps from Commander Second Army, 6:30 p.m., 11 April 1916.

51 Ibid., Second Army G.S. 617 to Canadian Corps, 12 April 1916.

52 LAC, RG 24, C-6-j, Vol. 6992, Vol. II, Chapter XI, "Canadian Corps, Reconnaissance of St. Eloi Craters and Position by Lieut. S.A. Vernon – Intelligence Corps. Night 11/12th April 1916," pp.1-2. After the battle, Turner noted, "Even Vernon, Corps Intelligence officer, went by compass and landed in craters held by our men," believing himself to have been in Craters 4 and 5. This clearly illustrates the difficulties encountered in navigating the battlefield at night, even with the aid of a compass. LAC, RG 9, III-D-1, Vol. 4679, Folder 14, File 1.

53 As Nicholson muses in the official history, "It is hard to understand why staff officers failed to interpret air photographs correctly. The photo taken on 8 April clearly shows Craters 6 and 7 half full of water and the others comparatively dry. Apparently no use was made of this obvious means of identifying the positions held by the Canadians." Nicholson, p.142.

54 Air photos taken at St. Eloi may be viewed at the National Archives of Canada in MG 30, E 46, Sir Richard Turner Papers, Vol. 2, File 10.

55 LAC, MG 30, E 241, D.E. Macintyre Papers, Vol. II, manuscript for *Men of Valour*, p.47.

56 Ibid., p.56.

57 One early example, titled "Notes Regarding the Interpretation of Aeroplane Photographs," was translated from the French and dated 20 October 1915. LAC, MG 30, E 61, C.H. Mitchell Papers, Vol. 16, File titled "Western Front Orders and Instructions 1914-1915."

58 The work was produced by Major J.T.C. Moore-Brabazon, a pioneer of aerial photography in the RFC. Anthony Clayton, *Forearmed, A History of the Intelligence Corps* (London: Brassey's, 1993), p.32.

59 H.A. Jones, *The War in the Air*, Vol. II (London: Hamish Hamilton, 1928), p.177; and S.F. Wise, *Canadian Airmen and the First World War, The Official History of the Royal Canadian Air Force Volume I* (Toronto: U of T Press, 1980), p.359.

60 Memoranda from April 20 and 24 illustrate in greater detail the new organization and procedure. LAC, RG 9, III-C-1, Vol. 3825, File "Can. Corps HQ-Files 10-13," Second Army, I.G. 924 (40/13), 20 April 1916, and Canadian Corps I.G. 999, 24 April 1916.

61 LAC, RG 24, C1, Vol. 20542, File 990.011 (D1); Typed excerpts from Haig Diary; diary entry for 17 April 1916.

62 LAC, RG 9, III-D-3, Vol. 5074, File 5, "Second Army," Canadian Corps G.220 to Second Army, 16 April 1916. Maurice Pope, a junior officer in the 4th Field Company, Canadian

Engineers at St. Eloi, remembered years later that, "One of the first actually to discover the mix-up in the craters to the left was…Lt. Bob Powell of the 4th Field Coy. who had vainly tried to decipher the puzzle from (our) first night onward. I so well remember him at CRE HQ 2 Div being shown air photos of the position by Bill Lindsay, the Chief Engineer [of the Canadian Corps], and his exclamation that at long last he could now realize what the position really was." Unfortunately, Pope does not provide a date for this story, but this statement may refer to the verification spoken of by Alderson in his 16 April memo to Second Army. DHH Kardex, 990.013 (D16), Collation of Comments, Colonel Nicholson's History of the First World War, p.230.

63 Cook, p.33.

64 Nicholson, p.144.

65 Ibid.

66 LAC, RG 9, III-D-3, Vol. 4843, 2nd Canadian Division General Staff War Diary, Appendix 183, Daily Intelligence Summary No. 205, 16 April 1916, p.2.

67 Cook, p.31.

68 LAC, RG 24, C-1, Vol. 20542, File 990.011 (D1); Typed excerpts from Haig Diary, diary entry for 21 April 1916.

69 Alderson continued his letter of 18 April 1916 with the following: "To myself I now put the matter this way:-

'Would you be as satisfied to place the 2nd Division in a difficult position as you would the 1st and 3rd Divisions'? The answer is 'No'. 'Why'? 'Because I have not the same confidence in the ability, the level headedness, the coolness and 'the thing has got to be done', spirit of the Commander'. 'Under these circumstances I feel that I have no alternative, sorry as I am to do so, but to submit that it is for consideration whether, or not, General Turner should remain in command of the 2nd Canadian Division.' I have personally communicated the above to General Turner and I am sending him a copy of this letter." LAC, RG 9, III-D-3, Vol. 5075, "Extracts from British War Diaries, Taken by D.H.S. July 1938," Canadian Corps G. 271 to Second Army, 18 April 1916.

70 Cook, p.32.

71 Turner to Duguid. No date (but probably datable to 23 August 1929, owing to a response from Duguid dated 16 September 1929 in which he remarks, "I have received your further comments on 'St. Eloi', contained in your letter of 23rd August…") LAC, RG 24, C-6-a, Vol. 1739, File DHS 3-17 (vol. 4).

72 General Sir C.H. Harington to Col. A.F. Duguid, 31 October 1929. LAC, RG 24, C-6-a, Vol. 1739, File DHS 3-17 (vol. 4).

73 Ibid.

74 LAC, RG 24, C-6-j, Vol. 6992, Chapter XI, Vol. II, "St. Eloi Craters (Appraisal).

75 Ibid., "Appraisal of St. Eloi Craters."

76 PRO, WO 158/296, Correspondence and Papers of Second Army Headquarters regarding St. Eloi, Canadian Corps G. 265 to Second Army, 18 April 1916.

77 Ibid.

78 For an examination of how the intelligence service within the Canadian Corps evolved, see Dan Jenkins, "Winning Trench Warfare: Battlefield Intelligence in the Canadian corps, 1914-1918," PhD Thesis, Carleton University, 1999. See also, Dan Jenkins, "The Other Side of the Hill: Combat Intelligence in the Canadian Corps, 1914-1918," *Canadian Military History*, 10, 2 (2001): 7-26.

CHAPTER 3

"A Bonny Fighter & a Born Leader"

A Portrait of Sir Archibald Cameron Macdonell, KCB, CMG, DSO

IAN MACPHERSON MCCULLOCH

> It is not sentiment, but simple fact, to say that he was loved and honoured – the grey old chieftain of the clans, a bonny fighter and a born leader.[1]
>
> G.G.D. Kilpatrick, Padre,
> 42nd CEF Battalion (Royal Highlanders of Canada)

Sir Archibald Cameron Macdonell, KCB, CMG, DSO, the first commander of 7th Canadian Infantry Brigade (7 CIB) and third commander of "The Old Red Patch" (1st Canadian Division) was unique, a throwback to the Highland chieftains of yore, and known politely amongst his peers and the officer corps as "Fighting Mac." The rank and file affectionately referred to him simply as "Batty Mac."

G.R. Stevens, who served as a private in the Princess Patricia's Canadian Light Infantry and was eventually commissioned in the regiment during the First World War, remembered Macdonell's forward style of command from the outset. "Batty Mac, our brigade commander, was crazy as a coot in many ways," he recalled. "I saw him actually get wounded one day. He was wearing...square-pushing jodhpurs...bright white...Somebody said 'Be careful, sir, there's a sniper' and he said '[Expletive deleted] the sniper', climbed up to get a look and the sniper took him through the shoulder and he went ass over apple carts into his shellhole from which he had emerged....My god, his language! You could hear him for miles around!"[2] The furious Macdonell got out of the shellhole and went back to retrieve his walking stick, whereupon another bullet passed through his left arm breaking the bone. Another officer who witnessed the episode recorded that "there is no doubt he would have been killed but for the fact the German sniper was so excited shooting at a General that he couldn't aim straight!"[3]

Brigadier-General Macdonell was perhaps one of the most eccentric, indomitable and beloved officers to have commanded troops in the First World War. Corporal Will Bird of the 42nd Battalion (Bn) remembered his brigade commander riding up to say farewell at Chateau de la Haie: "Everyone had a good word for him and stories of his decisions and actions were legion." But when it came time for Macdonell to speak "the old fire-eater seemed overcome with emotion...put his horse to the gallop and left without saying a word."[4]

Stevens described him in *A City Goes to War*, the Regimental history of the 49th Bn, as an officer of "a breed whose passing has left this world a poorer place – colourful, fearless, flamboyant in language, canny in battle, unabashedly sentimental and emotional over his men, who in turn regarded him with joy and pride; they treasured his eccentricities, they boasted (yes, and lied) concerning his highly individual behaviour."[5]

Archibald Cameron Macdonell came from a long tradition of military service[6] and started his own career by attending the Royal Military College (RMC) in Kingston Ontario where he excelled at games. At the end of his course at the College, his report from the examiner in Civil Engineering was, contrary to many academic reports, an accurate prediction. It read: "A man of marked ability who ought to rise rapidly in his profession and be an acquisition to any staff, on account of his high personal qualities."[7]

On 29 June 1886, Macdonell graduated and was awarded a commission in the Royal Artillery of the British Army. Due to financial difficulties in his family, he was unable to accept, but at once joined the Canadian Militia as a subaltern. Two years later, on 6 April 1888, he became a lieutenant in the Canadian Mounted Infantry of the Permanent Force (PF). The following year, however, saw him transfer into the Royal Northwest Mounted Police (RNWMP) in which he had a distinguished career for the next 18 years, counting service time in South Africa.[8]

Macdonell went to South Africa as a squadron commander in Lord Strathcona's Horse, winning a DSO and several mentions in dispatches before being severely wounded and subsequently evacuated back to Canada for a lengthy convalescence. Eager to get back, however, he was given command of the newly formed 5th Canadian Mounted Rifles (CMR) only to arrive in Cape Town a few days after hostilities had ceased. On return to Canada, Macdonell went back to the RNWMP but later transferred to Lord Strathcona's Horse. He was the regiment's commanding officer at the outset of war.[9]

An intelligent and worldly-wise man, Macdonell predicted in a letter dated 5 August 1914 to his nephew, Hugh Wallis, that "the war is likely to be a long one and many contingents will likely go, at least so it looks to me, & I firmly believe we shall win in the end, but Germany will take a lot of beating indeed." Wallis would become his uncle's Orderly Officer (OO) in 7 CIB HQ in December 1916, but not before joining the 16th Bn at Valcartier as a private, experiencing Second Ypres and serving as a platoon commander and scout officer with the 13th CEF Bn, (Royal Highlanders of Canada). Macdonell's prescient letter to his nephew, although unsuccessful in getting his headstrong relative to wait before enlisting, offers some interesting insights on his immediate prospects as a PF cavalry officer:

> They seem to be giving the cavalry the go bye altogether and I may have to accept an infantry command, a thing more or less incompatible to me. However in wartime, the unfortunate professional soldier must take what he is offered, one can't hang back, but I am much worried over the prospect. Colonel Steele considers that (as in the past) all plums will go to the eastern wirepullers.[10]

Macdonell went to the UK with the First Contingent, and later France, as the Commanding Officer of Lord Strathcona's Horse. After a brief stint as acting-commander of the Canadian Cavalry Brigade, Macdonell's pocket diary records on 17 December 1915: "Told officially I am to command 7th Infantry Brigade. Gazette to follow shortly." More interesting, however, are the rough notes in the back of the same diary, however, which outline his personal credo on how commanding officers are to command in his new brigade. They read in part:

> Orders - When you receive an order:
> Say what am I wanted to do? When
> am I intended to do it, then finish your appreciation.

1st eliminate self
Get ambition to do the best for [all]...
Read your orders carefully and say, Would I know what to do, etc.
Learn to check orders carefully with the map.

Points for COs
Care of men,
think of them all the time
How can they be kept dry, fed & spared
work, etc. shld be second nature

Discipline
Nothing without march
discipline '[illegible] keeps their place.
Inspection of arms shld be 2nd nature
Duty of an officer ditto in Infantrymen's feet.

HQs, Men & Horses
In battle habit is everything
Position of an O.C.'s [sub] units...
Make more use of horses and men mtd, etc they will help tremendously. CO should keep in close touch with General....officers should never be glued to comd HQs but try to anticipate events by being forward. Get the habit of frequent visits at regular intervals...[11]

The brigade commander refers to "habit" twice in his notes, stating that, "in battle, habit is everything". What does he mean? Macdonell, a veteran soldier, knew that battle was a chaotic affair and that in order to retain some semblance of control in combat, "standard operating procedures" (or "immediate action drills" in today's modern parlance) would provide the troops with a model or response to follow in times of extreme stress. Also of interest is his personal emphasis upon the importance of orders, his "directives" or operational decisions through which he would exercise command. He tells his COs he wants them to stop and analyse their mission first, do a time appreciation, then to formulate a plan to execute it thoughtfully and intelligently. He also stresses the administrative aspects of his commanding officers' responsibilities towards their troops and weapons in order to maintain his brigade operationally effective.

Finally, he was concerned with the passage of information and maintaining communications with his battalions. A cavalryman himself, and aware of the importance of timely intelligence and constant contact, he reminded his subordinates that they had horses. While he was not thinking of horses being useful in trench warfare, he was considering them no doubt as a means to speed up face-to-face contact while out of the line or during training. His command philosophy was clear in his closing remarks. He wanted his battalion commanders to be well forward and constantly visiting the men, anticipating their needs as well as the enemy's intentions and pending events.

While some of the above may have sounded like motherhood to the military man (two of his four COs were fellow regular officers, one Canadian, one British), Macdonell was impressing his personal style of command upon his subordinates and setting in place his command arrangements. He wanted to be perfectly clear on how the command process was to work within 7 CIB so that his "military Machine would run smoothly and well." And while the units he would "weld" together were undoubtedly "four of the finest battalions Canada had sent to the war," only one was a veteran battalion while the other three had no field experience whatsoever.[12]

When the decision was made to form the 3rd Canadian Division towards the end of December 1915, it was decided that the new division should be formed in France rather than in Canada or the UK. For its senior ranking brigade, it absorbed the Canadian Corps Troops Infantry Brigade (a two-day wonder) which was renamed 7 CIB on 22 December 1915. Macdonell's new command comprised the RCR commanded by Lieutenant-Colonel A.H. Macdonell, his cousin, the PPCLI commanded by Lieutenant-Colonel H.C. Buller, DSO, the 42nd CEF Bn (Royal Highlanders of Canada) commanded by Lieutenant-Colonel G.S. Cantlie, and the 49th CEF Bn (Edmonton Regiment) commanded by Lieutenant-Colonel W.A. Griesbach.

Of the four battalions, only the PPCLI were experienced veterans, having come from a year's hard fighting and distinguished service with the 80th Brigade in the 27th British Division. The RCR, at that time the only PF battalion in the Canadian Corps, had arrived in France in November 1915 after 11 months garrison duty in

Bermuda and had spent some time with 2 CIB doing work-up training followed by a stint of work parties. The two remaining units, the 42nd and the 49th, had both arrived in France one month earlier in October 1915, and had been placed under the aegis of 1st Canadian Division for trench warfare training. This consisted of no more than two 48 hour tours in the front line and the same period in reserve, followed by two months of relentless work as unofficial pioneer battalions.[13]

With the formation of 7 CIB, however, incessant labour behind the lines would cease and four very different infantry battalions would find themselves together for the duration of the war. When the PPCLI joined 7 CIB, they were equipped with short Lee-Enfields and Vickers machine guns, and, as the former "was universally coveted...all through the Canadian Corps, wise men of the Regiment did not leave their rifles lying about even in billets."[14] But the Regiment not only looked different outwardly. Internally, it was used to a different command system – the standard operating procedures of its previous brigade. Major Agar Adamson, the battalion Second-in-Command (2IC), was under no illusions that they had traded a comfortable, well-understood billet in a veteran, professional brigade only to join an "amiture [sic] army." He wrote to his wife two weeks after joining 7 CIB:

> I think both the division and brigade staffs are doing their best but they have had no experience and are in a shocking muddle. Orders are issued one moment and cancelled the next. We are in a constant state of change and jump, and so are they, nothing runs smoothly and none of the staff are sure of themselves. Very unlike the brigade they took us from where every officer had been a staff officer for years. I should not be surprised if they put us into the trenches and forget to relieve us.[15]

At that time, it was obvious that the Princess Patricias were the most experienced trench fighters in 7 CIB, let alone their division. Yet, curiously, there is no evidence whatsoever that their Brigade and Divisional commanders (or their respective staffs) recognized this fact by insisting they share their hard-won experience in helping train their sister battalions in the brigade or division. The staffs were too busy training themselves. Instead, the other infantry battalions would have to learn by trial and error.

Macdonell's forward presence, observation and supervisory abilities during this time are commented upon repeatedly in his subordinates' memoirs and letters. Lieutenant Royal Ewing of the 42nd wrote: "our Brigadier... is a corker – quite an old boy, but very active. He used to be up in the trenches at all hours of the day and night."[16] The 42nd War Diary (WD) was equally effusive: "During two months under the command of Brigadier-General Macdonell he had made himself respected for his tireless activity and much beloved for his interest in all of us."[17] Even the highly critical Major Adamson of the PPCLI was grudgingly forced to admit, though somewhat pompously: "I am quite pleased with our Brigadier, General MacDonald [sic]. He is always on the job and seems to know his job, and is most considerate....He can be depended upon to use good judgement and not rush into any sudden uncalled for move...."[18]

Whether Brigadier-General Macdonell's near fatal encounter with a German sniper illustrated good judgement is a matter for debate, but as a commander he was adamant that he would be well forward in the defence maintaining direct personal contact with his subordinates. This meant that his staff had to cope in his absence and there is some evidence that it struggled initially prior to the battle of Mount Sorrel.

The Brigade Major (BM) does not appear to have been as dynamic or keen as his commander on the "standing-up" of the Brigade, for as Macdonell noted dryly in his diary on 23 December 1915: "Major Foster, Royal Lancs, reported as Brigade Major and left at once on leave."[19] Lieutenant Hugh Wallis gives us a quick sketch of Foster's background when he wrote to his mother on 6 January 1916: "The Brigade Major has now arrived and seems a very decent and capable officer. He was at one time on the staff of the Lieut. Governor of India and was recently on the staff of one of Kitchener's divisions."[20] Foster would be wounded before the Mount Sorrel battle and temporarily replaced by Captain Cecil Critchley, Staff Capt "I", who would subsequently be wounded shortly after and replaced by Captain Basil Wedd of Toronto from HQ 3rd Division.

If Macdonell had no say in his choice of BM, he certainly could handpick the rest of his staff, a personal priority on learning he would command an infantry brigade. Lieutenant Wallis wrote home that "Uncle Archie has asked me to go to his Brigade as Orderly Officer, or what it amounts to, as A.D.C. [Aide de Camp] to himself." Wallis' initial perception of a brigade staff appointment was that it meant "a bed to sleep in, comparative safety, and many other advantages lucrative and otherwise," though this first impression would be rudely shattered as the year wore on. On 28 June 1916, the older and wiser Lieutenant Wallis would write, "The great objection I have to this place is that I get very little sleep whether there is a show on or not!"[21]

Captain Critchley, Brigadier-General Macdonell's PF adjutant from Lord Strathcona's Horse, would come over to fill in as the Staff Captain "I". That Critchley did not have any experience of intelligence work or how an infantry brigade operated was insignificant to Macdonell. Critchley could learn on the job. Macdonell's attitude was one that prevailed throughout the BEF with regard to professional intelligence work. "It was continually held [pre-1914] that the best man to help a commander assess the capabilities of enemy infantry was an infantryman and the best man to judge the potential threat of cavalry was a cavalryman," wrote the British Intelligence Corps historian. For an officer to devote his career to Intelligence was "in most Generals' opinion, a short-sighted policy which would lead to an officer having a specialized and narrow outlook to problems which required a wide and practical background of military experience."[22]

Lieutenant Wallis, as a former infantry battalion Scout Officer, became an indispensable assistant to Captain Critchley. "I am going to attach myself to Capt [Tom] Rush on the 'Q' side of the staff for the next few weeks as far as possible," Wallis wrote home, but the arrangement was only allowed by Critchley on the condition that Wallis "still take a hand in the 'I' side, paying particular attention to Sniping, Observation and Maps."[23] Captain Rush was another Strathcona import, having served as Macdonell's regimental Quartermaster (QM) prior to his staff job. Critchley would be promoted Major after Sanctuary Wood and would act as BM at the Somme battles when Wedd went out with pneumonia.[24]

Numerous references in Macdonell's diary indicate his preoccupation with finding a good Brigade Machine Gun Officer (BGMO), visiting no less a personage than Brigadier-General Harington at Corps to discuss the matter. Eventually his cousin

Archie Angus Macdonell, aka "Long Archie", commanding the RCR supplied him with the indomitable Captain H.T. Cock, the PF officer who had authored the CEF manual on the Colt HMG while serving in Bermuda. Lieutenant George Macdonald, PPCLI, whom Macdonell claimed as one of his own clan and affectionately addressed "Seorus Agraidh" ("Red George") in all correspondence, became the Brigade Grenade Officer (BGO), and later replaced Critchley as Staff Capt "I" when the latter was promoted and became BM.[25]

One area in which Macdonell had no initial say was in his battalion commanders. "Archie Angus" or "Long Archie", his cousin commanding the RCR, he knew intimately. Griesbach, a former CMR trooper in South Africa, was a cavalry officer acquaintance, as well as the son of a former RNWMP colleague. Lieutenant-Colonel Cantlie of the 42nd and Lieutenant-Colonel Buller, respectable officers and gentlemen both, were unknown qualities.

Griesbach was a man after Macdonell's own heart. Six days after the latter's return to HQ 7 CIB from sick leave in the spring, the Brigadier-General wrote to the CO of the Edmontonians and, in no uncertain terms, assessed his abilities:

> "A Good Commanding Officer makes a Good Regiment." Many have not the personality necessary at all to command men [underlining by Macdonell]. You have and I congratulate you on it. I should like to congratulate you also in the straight and fearless way in which you have reported on the Ross rifle. If only all COs would write or speak out as straight (not only about the rifle; also about all things concerning their commands) smooth working and officering would be largely increased.[26]

Arriving back from his convalescence in hospital, Brigadier-General Macdonell must have also taken stock of the new CO of the RCR who had replaced his cousin in April when the latter left to take command of the 5 CIB. Lieutenant-Colonel Claude Hill, a permanent force (PF) officer of 15 years service prior to the War, came to the brigade with 18 months experience as Bn 2IC of the 24th Bn, Victoria Rifles of Canada (VRC), in 2nd Division. It was Hill who would earn the nickname "The Shino Boys" for the RCR the day after he took command when it came out of the line in the salient on 21 April. The regiment was greeted with the new CO's orders "that all kit must be cleaned and all buttons kept shined when the unit was in billets behind the front line. Some grumbling resulted; but the Permanent Force officers, with a strong belief in the efficacy of smartness in enhancing morale, welcomed the orders and supported the Commanding Officer's stand with a firmness that soon ended all opposition."[27]

Whenever Macdonell visited the battalions or higher HQs, he always took a staff officer with him, usually his nephew Hugh Wallis, and later in his tenure, his other nephew, Harold. Hugh's brother, Captain Harold Wallis of the 16th Bn, came to 7 CIB in October to replace his brother as OO. Hugh Wallis had moved up to Staff Captain "I" in September replacing George Macdonald, wounded at the Somme. The flexibility of brigade staff officers is shown when he wrote his mother: "Capt [P.E.] Coleman has been away on leave so I have been looking after A&Q [Administration & Quarters] while Harold has been on "I" for me. I am OC Rear HQ consisting of the interpreter and myself; [the paymaster] is on leave and the veterinary officer at division, so I have my hands full." Two weeks later he wrote "this short tour in charge of A&Q of the Brigade has given me a great chance to learn things which I have not had an opportunity of doing before."[28]

That Macdonell allowed his fledgling staff some latitude within the scope of their responsibilities is evidenced by remarks made by Wallis during the temporary command of 7 CIB by F.O.W. Loomis (14 March - 5 May), standing in for Macdonell while he recovered from his wounds. After serving only 12 days under the newly-promoted Brigadier-General Loomis, Lieutenant Wallis was writing home: "I hope Uncle Archie will be back soon; otherwise I shall resign;...F.O.W.L. my old battalion O.C. is in command here now." Three weeks later, he was more emphatic about Macdonell's return:

> Nothing will please me so much as Uncle Archie's return; everyone is just sort of standing by until he comes back. I don't seem able to do a thing under the present regime although the G.O.C. [General Officer Commanding] is very decent to me, everyone has difficulty carrying on because of his perpetual interfering and directing. I often feel like asking to go back to the regiment, but must wait and have all the information I can, to give to Uncle Archie.[29]

The Loomis "regime" serves to highlight Macdonell's personal command style and philosophy. Loomis's apparent over-supervision trampled the trust and confidence that had been slowly building up within the HQ staff as they learned their trade, and it is not apparent from unit war dairies, regimental histories or letters that the stand-in brigadier ever visited the forward battalions, thus making his presence felt to the troops under his command. Hearing of his HQ's malaise via various visitors to his convalescent hospital, Macdonell reappeared on 6 May 1916, before his wounds had truly healed, the doughty warrior convinced his clansmen needed him. A relapse in the third week sent him back for several days, causing Lieutenant-Colonel Buller, PPCLI, the senior officer commanding, to move across as acting-brigade commander, but at least Loomis had moved on to command of the 1st Canadian Division.

The three major engagements of the Canadian Corps in 1916 – the St. Eloi Craters fiasco, Mount Sorrel and the Somme battles – reflected the ongoing difficulties of overcoming the deadlock of trench warfare. "Deft coordination of all the components of the Corps, careful preparation prior to battle and skilled use of artillery resources that would characterize the Canadian performance in 1917 and 1918 were absent," noted William F. Stewart in his contemporary assessment of CEF performance.[30] Some progress had been made towards fielding a more decisively lethal force, but too many officers and men still had to learn their jobs. Not only was the staff of 7 CIB raw in many ways, its parent formation, 3rd Canadian Division, was only activated in the winter of 1915/16 and declared operationally ready as a division in March 1916. An even more woeful novice was the 4th Canadian Division, which landed in France in mid-August 1916 only to be committed to battle the following month. Of the three "shows" of 1916 mentioned above, 7 CIB found itself deeply involved in the latter two actions, the first one defensive, the second, offensive.

The defensive battle, Mount Sorrel, "constituted only a small, localized operation of little or no significance to the outcome of the war" according to historian D.J. Goodspeed, "...in comparison with the monstrous battles of Verdun, the Somme and Passchendaele..."[31] Yet to the 8430 Canadian soldiers who would be killed and wounded in the twelve day clash, it was anything but a "small localized operation." The Imperial German Army launched a massive spoiling attack on the salient in the hopes of tying down Entente forces and preventing their transfer to Picardy for the pending Somme offensive. In the end the strategic aim of the German operation was not achieved, and tactically the Germans could claim little better than a draw for the original front was established in most of the sector.[32]

For Brigadier-General Macdonell and 7 CIB, Mount Sorrel would be their first major combat operation as a brigade. "The June Show", as it came to be called, occurred at the apex of the Ypres salient, a comparatively flat piece of terrain, dominated in the south-easterly portion by a low wooded hill known as Mount Sorrel. Practically all the ground held by 3rd Division in the forward area was still wooded at this stage in the war, its frontage of 2500 yards running from the Menin Road in the north, south to Mount Sorrel inclusive. As a control measure, this in turn was divided into two brigade frontages of which 7 CIB had responsibility for the left and the 8 CIB the right on the day of battle. Each brigade defensive scheme called for two battalions in the front line, one in support and one in brigade reserve. The 7 CIB's two front line battalion sub-sectors were known as the Hooge sector on the left, the line running through the ruins of the village of that name, and the Sanctuary Wood sector on the right. Both of these sectors were dominated by Mount Sorrel and two other hills to their south in the 8th CIB sector. Thus the PPCLI History notes: "It was early recognized that if the Germans made a determined attempt to reach Mt Sorrel, this front line would be quickly blown out of existence. The defence scheme for Sanctuary Wood and Hooge was therefore dependent upon resistance which might be provided by the second or R-line."[33]

On the morning of 2 June 1916, German artillery fire on the 3rd Division's area developed into the heaviest bombardment endured by British troops up to that stage in the war. Trenches and their garrisons holding them vanished in clouds of dirt and dust, while whole trees in Sanctuary Wood were hurled skyward by the bursting shells. After exploding four large mines slightly forward of Mount Sorrel, the German infantry overwhelmed the 8 CIB front line trenches and captured the important high ground as well as Hills 61 and 62 by nightfall.

The PPCLI, as right forward battalion in the Sanctuary Wood sector, was on the northern flank of the main German assault. Its right forward rifle company was virtually annihilated at the outset by artillery fire but its left forward company shifted left to help the survivors and together they grimly fought back. Nicholson's Official History rightly states: "Credit for temporarily checking the enemy's right wing belongs to the [forward] Patricias...As the Germans surged eastwards, its rifles volleyed into the enemy's right rear. [They were] to hold out successfully for eighteen hours, isolated from the rest of the battalions and with all their officers killed or wounded."[34] The PPCLI CO was killed leading the remnants of his two other rifle companies in repeated counter-attacks up and down the support and communication trenches of 8 CIB during the afternoon, buying valuable time for the RCR support companies and two 42nd Bn rifle companies trying to shore up the R-line 500 yards to the rear of the PPCLI. Two other 42nd rifle companies were sent over the brigade boundary into Maple Copse at the rear of 8 CIB's collapsed front line in order to prevent the brigade being flanked from that direction.

In essence, the first German success at Mount Sorrel had obliterated all brigade boundaries, and the most desperate and important fighting of 7 CIB – the defence of Warrington Avenue and Lover's Walk by the PPCLI on the afternoon of 2 June, and the counter-attack of the 49th Battalion the next day – took place largely on 8 CIB territory.

The RCR, the left forward battalion in the Hooge sector had minimal activity on its frontage but soon had to turn its right flank towards the enemy once the survivors of the PPCLI front line companies withdrew to the R-Line. Brigadier-General Macdonell liaised with the British formation on his left flank and arranged for British troops to take over the RCR's responsibilities beyond Hooge so they in turn could shift right. HQ 3rd Division was fighting its first defensive battle without the benefit of a commander. Major-General M.S. Mercer had been killed by German artillery while on reconnaissance in the 8 CIB sector, and the latter formation's commander wounded and captured.

Once HQ 3rd Canadian Division had confirmed that the 8 CIB frontage had been indeed captured, it ordered 7 CIB at 5:30 pm, 2 June, to restore the situation, giving Brigadier-General Macdonell, two battalions from its reserve brigade, 9 CIB. Corps intervention later revised the initial plan and set the time for a coordinated divisional counter attack, reinforced by 1st Division units, for 2:00 am the following day. An immediate counter-attack by 7 CIB, however, was virtually impossible, whoever was ordering it. The brigade's reserve battalion, the 49th, which had stood to since 10:00 am, 2 June, was far to the rear in Ypres and had to come forward through smashed-in trenches and an unceasing hail of high explosive, shrapnel, and tear gas shells, as did the two 9 CIB battalions which were still further to the rear.[35]

Brigadier-General Macdonell ordered the 49th CO to physically coordinate and command the attack on the ground forward, but due to darkness, confusion wrought by the heavy bombardment and lack of communications, the 2:00 am "Zero Hour" for the attack was repeatedly delayed until 7:00 am the following day. The 7 CIB attack did not go in with the others on their right flank as two 9 CIB battalions failed to show up at the assembly area on time. Finally, an exasperated Lieutenant-Colonel Griesbach ordered the only battalion in position, his beloved 49th, to attack alone in broad daylight at approximately 9:00 am. With virtually no artillery preparation, the Edmontonians suffered 358 casualties in the space of a few short minutes, achieving nothing.[36] The 42nd History described it as a "forlorn hope," as no other supporting attacks on the right from 1st Division accompanied it, thus allowing the Germans to concentrate maximum firepower on the lone battalion advancing in the open.[37] The 49th's losses on 3 June were only surpassed by the veteran PPCLI battalion who had suffered over 400 casualties, though these were sustained over a 24-hour period rather than in a mere 20 minute time span.

With two battalions severely mauled, senior commanders killed, and a failed counter-attack to its credit, the battered 7 CIB was finally relieved on 5 June 1916. Less than two and half months later, however, they were sufficiently rested and reorganized, given some basic assault training, and sent to take their turn in the "meat grinder" engagements further south, known collectively as the Somme battles.

During the first of the two major operations of 1916, the battle of Mount Sorrel, the picture that emerges of Macdonell is of a cool, calm, collected commander, working behind the scenes to ensure a coordinated effort and trying make sure his troops were in the right place at the right time with the right resources. In occupying the northern half of the Salient he had always been concerned about the high ground on his right flank to the south where even the most limited penetration could bring the enemy into his right forward battalion's rear. However, through his foresight and planning he had sought permission from 8 CIB when they had occupied their sector prior to the battle, to position No. 3 Company PPCLI in that brigade's rear area in case of such an eventuality. He and Lieutenant-Colonel Buller had readily appreciated that if the enemy were to gain Warrington Avenue, a communications trench which angled back from the 1st CMR's front to his reserve line, his entire brigade position would be compromised.[38]

In essence, his brigade "vital ground" (ground, which if lost, renders the commander's position untenable) was in another brigade's area of responsibility. By wisely anticipating the enemy's possible actions through a worst possible scenario approach, Macdonell ensured before the battle ever began that he had balance – a better chance of defending his sector than if he had slavishly adhered to brigade boundary lines on a map. One could readily agree with a brigadier commenting in 1917 that "the battle 'command' is today subordinated to the battle 'preparation'. Our fights are won or lost before we go into them."[39] Though Victor Odlum was referring to the set-piece assault, his comments certainly have some validity for the British trench defences of 1916, especially those in the Ypres salient where the enemy was consistently blessed with the initiative and dominating ground. How the defensive battle was to be fought had to be well thought out beforehand and discussed and rehearsed from the battalion level of command right down to the platoon. Each battalion commander was responsible for having his own battalion defensive scheme.

During the 2-3 June battle, command and control at the brigade level of front line troops was, for all intents and purposes, completely lost as all communications were cut by shellfire except for the RCR on the extreme left. Lieutenant-Colonel Griesbach wrote after the battle that he "could not conceive how higher command can influence the defence of positions without some better means of communication than now exist. As it stands, an attack might engulf the whole of the frontline troops, and the fugitives' arrival at the Brigade HQ might be the first intimation of the fact."[40] But, as most enemy assaults were accompanied by a massive barrage, defenders usually had more than adequate warning.

When it was obvious the German artillery barrage of 2 June was abnormally heavy and not abating, Brigadier-General Macdonell had ordered the 49th to "stand-to" at 10:00 am and be ready to go forward. He did not need orders from higher command to understand the essentials of his mission. It was a standing order to hold the front line at all costs, a task which his veteran Patricias, bearing the brunt of the attack that day, knew only too well. The regiment's steadfast performance and delaying actions gave Macdonell the requisite time to shift his forces to consolidate the R-line and to move additional troops to the rear of 8 CIB. Forward of the R-line, the battle became very much a company and battalion commanders' "show", local initiatives and common sense dictating the defence.

The only failure in Macdonell's conduct of the brigade defensive battle could be said to been the abortive counter-attack of the 49th. Once the orders were issued for this attack, Macdonell had no control over the proceedings, having delegated authority for its ultimate conduct and launching to Lieutenant-Colonel Griesbach, his designated commander on the ground. However, like a true commander, he recognized that delegation of the task did not necessarily absolve him of the responsibility for its failure and that he must ultimately share some of the guilt as he had delegated a task to a subordinate without fully ensuring that subordinate had the requisite resources or time to accomplish the task, in this instance — adequate artillery fire support and the additional manpower of two 9 CIB battalions placed under his command for the operation. After the battle, Macdonell shielded Lieutenant-Colonel Griesbach's actions in ordering his sole counter-attack battalion forward in the doomed assault (a decision which must have taken a great deal of soul-searching on Griesbach's part, as it was his own command) and took full blame for all mistakes upon himself in his official report to higher HQ stating:

> In reviewing the work done by the 7th Brigade, perhaps the proudest thing I can say is 'the machine worked smoothly and well.' It follows that the mistakes — and what military operation takes place without them? — were my own. Let the splendidly gallant officers and men who carried out my orders faithfully unto death and held the single line trench at that time was our only bulwark against defeat, receive their measure of earned praise full to overflowing.[41]

This propensity of Macdonell for downwards loyalty secured for him the love and respect of his subordinates. One battalion commander told Macdonell's biographer, A.E. Kennedy-Carefoot, that "in his early army history when he was "on the mat" before his superiors, Macdonell [had] stood by him splendidly. Such loyalty to one of his subordinates was unforgettably appreciated by the hapless victim." Griesbach, in an interview with Sir Arthur Currie's biographer, Hugh Urquhart, after the war, acknowledged that the officers and men under Macdonell's command, "could only be attacked through him, which is alright up to a point, but which has its disadvantages. In short, I would say that with General Mac it was a religion to stand up for his subordinates on every occasion."[42]

In retrospect, Brigadier-General Macdonell could have told higher HQs before his brigade counter-attack went in, that their timing for an immediate counter-attack had long since passed and that he had not been allocated nearly enough time for what was really the mounting of a deliberate counter-attack. Currie commanding 1st Division certainly did, but it didn't help. When the Corps order was issued at 10:15 pm in the evening of 2 June, several factors dictated that the latter form of assault was the type actually required: the Germans had already been consolidating their gains around Mount Sorrel for some six hours; the troops who would have to be used for the counter-attack were several miles from their projected forming-up places and would have no time for reconnaissance; artillery fire plans would have to be prepared without accurate information on the new enemy positions; and nothing had been done to silence or neutralize the enemy artillery which was still in great preponderance around the salient.

In short, every basic principle for ensuring a successful deliberate counter-attack would be ignored under the pretence that an immediate counter-attack was being mounted and these oversights could therefore be excused. Historian D.J. Goodspeed noted, "in light of all this it is difficult not to regard General Byng's order as an emotional rather than a rational reaction. Whatever the reasons for the order were, it was, as always, the troops who had to pay for the mistake. In the event, and not surprisingly, almost everything that could go wrong did go wrong."[43]

Whether Macdonell actually believed the ordered counter-attack could succeed is not recorded, but at least the PPCLI and 49th Battalion histories noted his more immediate concern as being the PPCLI's precarious state, verging on breakdown. "Not to attack," recorded the 49th History, "meant leaving the Patricias to their fate." The death of the PPCLI's CO, the serious wounding of their 2IC, and heavy casualties among the senior officers and senior NCOs, had left command and control in the hands of a few junior officers and NCOs, most of whom were also wounded. The PPCLI History confirms in a rather convoluted fashion that "the Patricias in the line had borne the strain of twenty-four hours of constant bombardment and fighting superbly; but though they were still 'perfectly cheerful' they were becoming physically and nervously exhausted."[44]

Because Macdonell stood by him, Griesbach survived, and his vindication as a good CO was reflected in his solid performance in future battles and subsequent promotion to Brigadier the following year. Griesbach knew whom to thank, telling Macdonell in a farewell letter that "Your treatment of those under you has had a result which you may not have foreseen – your Commanders have always been free men – free to serve you without fear of anything underhanded. Proud to have your commendation and too proud to merit your censure."

Another letter after the war, (Griesbach having served yet a second time under Macdonell in 1st Division), would be even more frank, stating: "To you I owe more than I can say. Friendly and constructive criticism, sound advice, generous encouragement and appreciation — all these and more I have received at your hands. I have not only had justice from you but also mercy, and at times, was more in need of mercy than justice."[45]

After Mount Sorrel, the Canadian Corps moved to the Somme, that archetypical First World War battle which saw long linear waves of heavily burdened infantrymen plodding forward shoulder to shoulder as Maxims mowed them down. It was already a killing ground when the Canadians arrived in September, while Macdonell was trying to cope with a deep dark depression brought on by the news of the death of his only son, Ian, flying with the RFC. Distraught and distracted, he felt he could not devote his full and proper attention to the pending attack. According to his biographer, he went to HQ 3rd Division and had a personal interview with Major-General Louis Lipsett. "He asked to be excused from the attack." Lipsett expressed sympathy but told him it was out of the question and that he must command the attack.[46]

The 3rd Division, containing the 7 CIB, was to be employed both in September and October on the narrow front bounded on the north by the winding Ancre River and on the south by the line of the Albert-Bapaume road, which bisected the town of Courcelette. As the left forward division of the Canadian Corps, 3rd Division had the role of facing Ancre Heights to the north and securing a defensive left flank to the general Corps attack being pressed forward astride the Albert-Bapaume road. On 15 September, the Canadian Corps was ordered to attack on a two division frontage, 3rd Division left forward with one brigade up (8 CIB) to provide flank protection, and 2nd Division right forward to attack the forward defences in front of Courcelette. The 7 CIB was called up to exploit the success of 2nd Division in front of Courcelette pushing in between 8 CIB on its left and 5 CIB on its right. The 7 CIB secured the FABECK GRABEN Trench with the 42nd Bn and the PPCLI with minimal trouble and casualties thereby securing 5 CIB's left flank as it captured Courcelette with relative ease.[47]

The 7 CIB's luck ran out the very next day, however, as German defenders rallied and stiffened their defences with a fresh brigade of battle-hardened Marines. The RCR and 42nd battalions going forward early the next morning with little artillery preparation, hoped to exploit to the next line of trenches, but suffered heavy casualties as a result of their unimaginative effort. Taken out of the line to refit, the 7 CIB would be committed to battle the following month in 3rd Division's continuing futile attempts to take the infamous Regina trench. This time, the 49th Battalion and the still weakened RCR would make disastrous unsupported frontal attacks, and although the RCR would break into the trench system, it would be repulsed by repeated counter-attacks and sustain over 70 per cent casualties.

The attacks that the 7 CIB were called upon to mount at the Somme have been termed "deliberate" though in fact they were poorly coordinated "hasty" attacks. Major-General Lipsett himself recognised that Macdonell's men's achievements on the 15 September 1916 were essentially the combination of good luck and good reconnaissance work. He wrote afterwards:

> The problem which faced the commander of the 7th Brigade was a hard one. Four and a half hours only were available to march five miles over difficult country devoid of landmarks, through enemy barrages, to deploy

> for attack in broad daylight in a captured and partially obliterated German trench, the whereabouts of which was not known, except from the map, to the battalion commanders, and to launch the attack, on a two-battalion front at 6 P.M. Nevertheless, owing to the previous excellent reconnaissance work of the regimental scout officers and scouts who had been sent ahead, and who met their battalions en route, all battalions were in their places on time, and the attack went forward punctually.[48]

That luck played a part in the afternoon's proceedings is underlined by the following revelation in the 42nd Battalion history: "No detailed battalion operation order was issued owing to lack of time and so hurriedly was the whole undertaking carried out that it is doubtful whether anyone, except the officers and a few of the NCOs, had any definite understanding of the exact task and even these necessarily had but slender knowledge of it."[49] The Brigade was thus launched and Macdonell's command and control capabilities with regards to the attacking battalions thus ended. The onus fell upon the battalion commanders to carry the attack forward.

Macdonell still retained command and control of his support and reserve battalions however, units which he could feed into the battle as he saw fit and thus still influence the outcome. He also controlled the reserve elements of the Brigade Machine Gun (MG) Company (Coy) and Trench Mortar (TM) Battery. The battalion commanders, to a degree as well, became impotent as their assault companies went forward from their designated jumping off lines. The CO usually stayed back with the reserve company and HMGs or in a nearby dugout. The Battalion commander's role in battle was to act as a sort of advanced report centre for his forward companies to report back to so information could be relayed back to the Brigadier or laterally to flanking units. He too could manoeuvre his reserve and HMGs, request reinforcements or artillery support, liaise laterally with flanking units, or order withdrawals or realignments as required.

Despite the 7 CIB's problems on the Somme (most not of its own making), the second year of the brigade's existence under Macdonell was one marked by a higher level of competence and professionalism, higher standards of training, reorganization of its fighting components at the brigade, battalion, company and platoon levels and the development of a distinctly Canadian attack doctrine. New applications of technology and tactics included the massive HMG barrages and interdiction introduced on a grand scale at Vimy, counter bombardment and sound-ranging techniques, counter-electronic warfare, the introduction of new gases, aerial photography, increased use of wireless, the wide spread use of the grid communications system, and fire and movement at the platoon and company level. In the course of the year, the 7 CIB would participate in two major operations, Vimy and Passchendaele, and a brigade-sized raid conducted at Avion in June 1917. After the immensely successful Vimy battle, Macdonell would be earmarked for higher command.

Interestingly, before Vimy, Byng, as a new corps commander and a stranger to Canadians, had sought out a professional cavalry officer like himself for a personal heart to heart discussion on the calibre of Canadian platoon officers arriving in theatre. The acting-BM of 7 CIB, Captain Cecil Critchley, recorded in his memoirs that Byng "came to visit General Mac, whose office was next to mine [and] as there was only a curtain of sacking between us, I could hear what General Byng said." The two generals' conversation went something like this:

> "I am not satisfied with the training Canadian troops are getting in England, Macdonell. The officers seem to come out with no practical training at all. Officers and men very often get killed quite unnecessarily, just through lack of proper knowledge. You, General, have had considerable experience of the Canadian Forces. You have been with them a long time. You know a great number of officers. I've really come to talk the problem over with you."
>
> "First of all, I must develop a training centre out here, both for those who come out as reinforcements, or people I want to send back from the front line to get further instruction. I am therefore going to establish a corps school."
>
> General Mac said: "I think we better have Critchley in. He knows most of the officers in the Canadian Corps and his advice might be useful."[50]

In the conversation that followed, Byng interrogated the 26-year-old Critchley on suitable candidates to run the new Canadian Corps Training School (CCTS) he was proposing. After some discussion with Macdonell and his BM, General Byng "looked up from under his shaggy eyebrows" at Critchley.[51]

> "What about yourself? Could you do it?"
>
> I said, "I beg your pardon, Sir, I am acting brigade major here."
>
> He knew as well as I did that the brigade major's post was a considerable plum for a regular officer, but he said sharply: "You will do as you are told. Come and see me in three days' time and I'll find out more about you. Is that all right, Macdonell?"

Brigadier-General Macdonell did not have much say in the matter and lost his acting-BM by the end of the week. Byng, satisfied with his inquiries regarding Captain Critchley, made him the new General Staff Officer 2 (GSO2) (Training) on the Canadian Corps HQ staff. Critchley would go on to establish and command the Canadian Officers Training School at Bexhill in the UK and, by the end of the war, would be seconded to the RFC/RAF to revamp their entire training establishment. By September 1918, Critchley was a 28-year-old Brigadier-General.[52]

The story of the Canadian Corps' 9-12 April assaults on Vimy Ridge is perhaps the most well-documented in Canadian military history and will not be gone into any great detail other than to outline those salient points required to provide some historical context for the operations of 7 CIB before, during and after this seminal battle. Lieutenant-General Byng's plan called for a frontal attack by all four divisions of the Corps in numerical order from the right to left. The 3rd Division, facing La Folie Wood, was ordered to attack on a two brigade front. The 8 CIB would be on the right and the 7 CIB on the left, the latter brigade advancing in the shadow of Hill 145.[53] In the 7 CIB, three battalions would take part in the assault, each advancing with two companies up and two in support ready to pass through on the first objective line: the RCR right forward; PPCLI centre; and 42nd Battalion left forward. The 49th Battalion would provide mopping-up and carrying parties for the assault battalions.[54]

As a whole, the attack of the Corps was to be carried out in four stages, each dictated by the German zones of defence. On 3rd Division's frontage, the operation would only entail participation in the first two stages: an advance at 5:30 am, scheduled to reach the first objective 35 minutes later. After a 40 minute pause to reorganize, a subsequent advance would be made at 6:45 am to a line drawn through La Folie Wood, along the reverse slope of the Ridge and bending back on the left to conform with the objectives of 4th Division. The 7 CIB was allocated 20 minutes for this second phase. The 1st and 2nd Divisions had objectives at a maximum distance of 4000 yards from their jumping off positions while the 4th Division had the shortest distance of all but was faced with the prospect of seizing Hill 145, the strongest natural defensive position on the whole front.[55]

It should be noted here that, German defensive lines and tactics had changed significantly everywhere, except on the Ridge allocated as the Canadians' objective. After the Somme, the German army had gone to a more "elastic" defence-in-depth by zones, but on the Ridge their dispositions were a hold-over from the past, "designed for Falkenhayn's stonewall tactics, which proved so disastrous on the Somme," notes Byng's biographer, Jeffery Williams. "A more elastic system was planned but work on it had not begun. In the face of increasing destruction caused by the British and Canadian guns and frequent alarms caused by infantry raids, the garrison could do little more than maintain their existing defences."[56]

At 5:30 on the morning of 9 April 1917, the rolling barrages opened, and the attacking waves of the Canadian divisions went forward. Waves, in the case of 7 CIB, however, are not quite an accurate description. All of its assault battalions had to cross a series of large craters to their front before seizing their first objective line. It took the 42nd on the left five minutes to "scramble across the muddy craters as best they could, then re-form with great steadiness just as though they were rehearsing over the tapes at Bruay."[57]

By 7:30 am, all three assault battalions were on their final objectives mopping up, but the morning was just beginning for the brigade's left forward battalion. The 42nd Royal Highlanders from Montreal had to re-group in the shadow of Hill 145, the highest and most important feature of the Ridge. The defences of this hill were particularly strong, ringed with well-wired trenches and a series of deep dugouts on its rear slopes.

The 42nd got to its final objective line initially unhindered because these German defences were being heavily shelled. Once the Canadian artillery lifted, Hill 145's defenders were too preoccupied in bloodily repulsing 11 CIB's attempts to seize it. When 11 CIB went to ground short of its crucial objective however, its defenders had more time to take stock of their situation and see that they could enfilade 7 CIB's position with relative ease in La Folie Wood on their left. The majority of the 42nd Bn's casualties sustained at Vimy were thus incurred by sniper, MG and observed artillery fire being brought to bear from Hill 145 on their positions in and around their consolidated position on the final objective.[58]

The major problem of command and control during the battle for Brigadier-General Macdonell would therefore be one of liaison over divisional boundaries with a flanking brigade. His immediate concern was to consolidate his own defence and to provide what assistance he could the following day for the capture of the troublesome hill on his left flank. Two days after the capture of Hill 145, 10 CIB would successfully storm The Pimple, by which time, the enemy accepting the loss of Vimy Ridge as permanent, had pulled back two miles to their Third line in the new Hindenburg system running southeast from Lens across the open plain.

"The great lesson to be learned from these operations," read one divisional after-action report, "is this: If the lessons of the war have been thoroughly mastered; if the artillery preparations and the support are good; if our Intelligence is properly appreciated; there is no position that cannot be wrested from the enemy by well-disciplined, well-trained and well-led troops attacking on a sound plan."[59] After Vimy, Brigadier-General Macdonell, in his familiar role as keeper of "The Fighting Seventh"'s *esprit de corps*, noted in his congratulatory address to the troops that the key factor in their success had been time. "Never before," he rightly stated, "had we the chance to work up to an attack in detail...Our training was done with thoroughness and proved of incalculable value."[60]

Comprehensive, progressive, and realistic unit and formation collective training was backed up with an extensive training system featuring individual, specialist and reinforcement training behind the lines. The move to the Vimy Front had been accompanied by the establishment of the CCTS under the command of Major A.C. Critchley, former BM of 7 CIB. The CCTS was divided into an Officers'

Wing and an NCO Wing to provide instructors for the Divisional Schools. With its own training establishment in place, the Corps was well on the way to ensuring its own particular doctrine was well disseminated and understood by all ranks. But as 7 CIB would learn at Avion in early July, the tactics required to take German defences at Vimy, based to a large extent on the old pre-Somme model, were not necessarily applicable to the elastic defence system of pillboxes they would encounter at Passchendaele or the next year in their assaults to break the Hindenburg Line.

Initially, the Avion Raid was to have been part of an limited offensive to "inch up" on the city of Lens. Currie, short of supporting guns in early June and thus unable to prevent the enemy from concentrating overwhelming fire on any trenches he might seize, convinced Haig that trying to hold captured ground at great cost would be unproductive at that time.

Orders called for six battalions to attack on the night of 8/9 June at midnight: three from the 11 CIB attacking to the north in the La Coulotte sector (out of contact with 7 CIB), and three from 7 CIB in the Avion sector with 49th Bn left, 42nd Bn centre and the RCR right. Each battalion of 7 CIB would go in on a four-company front, each with three platoons in successive waves, the fourth platoons of each company staying to constitute the garrison of the jumping-off trench. The 7 CIB battalions were to attack on approximately 1200 yards of the German front in the Avion area, penetrate into enemy lines to a maximum depth of 800 yards, hold the ground captured for an hour and a half, then begin a covered withdrawal, completing the evacuation in a 30-minute time span.[61]

Preparations for the raid, which would take place at night, were conducted with Vimy-like thoroughness behind the lines. Rehearsals were conducted by day and night over taped ground and troops instructed to memorize the exact locations of all known enemy MG posts, trench mortar posts, and dugouts. The composition and duties of trench raiding parties received particular attention.[62]

In the aftermath of the raid, the newly promoted Major-General Macdonell, of course, regarded it as a huge success. "Fighting Mac" wrote to a subordinate officer two days later: "The Brigade put on a good show the other night and we are beginning to consider that we are 'Sturm Truppen' for fair. It was a good 'Show', well carried out and successful in every respect."[63]

By contrast, the view from the ranks was considerably different, especially in the 49th Bn, which sustained the highest casualties going in on the left flank of the raid. From Sergeant A. Fowlie's perspective, whose platoon was virtually wiped out, "it was a very fierce affair and it lacked the luck. We held the trench and the Germans infiltrated my company...I lost three very good friends of mine in that raid. I don't think it was a very good planned affair after all."[64] On withdrawing after two hours in the enemy's trenches, the three battalions forsook the cover of the German trenches to move back across No Man's Land and were caught in a heavy German counter-bombardment. In the words of Lieutenant G.D. Kinnaird, 49th Bn, who was severely wounded, "the moppers-up were mopped up."[65]

The 42nd Battalion history deemed the raid to be "the most thoroughly organized and brilliantly carried out minor operation in which the Brigade ever participated as a unit." As a battalion, the 42nd went in with its flanks secured by the other two battalions, though it had the second highest rate of casualties. It went on to claim, somewhat correctly, that the raid was out of the ordinary in that "it was carried out on a very much larger scale than had ever been attempted" with the added difficulty of having been conducted at night.

7 CIB reported that it had sustained 335 casualties in the raid, 38 of them fatal, but a quick tally of the battalion counts show the actual killed total to have been 44 all ranks.[66] Quick to justify the raid in terms of its value, 7 CIB claimed an inordinately high body count of 560 enemy killed, an entire battalion's worth. This figure is not realistic as the Germans only manned a regimental sector with one battalion in the first two lines of trenches and many of the frontline and second line defenders retreated to their third line during the attack. Many of the dugouts destroyed by the 7 CIB raiding force would thus have been empty.

The 49th Regimental historian asks a valid question and leaves the answer unstated, but obvious: "Did this operation yield a credit balance?...In the 49th records one account declares that the attack went in perfect alignment; another, that it was all confusion.... The menace of the counter-barrage perhaps had been over-discounted.

In the darkness the man who sits tight with his weapon is certain to enjoy an advantage over his adversary who stumbles across open ground to seek and destroy him."[67]

The planning and training for the Avion Raid was "Fighting Mac's" last "show" as brigade commander, the actual raid taking place after he had left to take up command of 1st Division. At home in Canada, his promotion to Major-General had been questioned by Borden who tried to block it on the instigation of Major-General Sam Steele, the latter feeling that the command should go to Sam Hughes' son, Garnet. The new corps commander adamantly refused to have Hughes in lieu of Macdonell. George Perley, the Minister responsible for Canada's military forces overseas supported Currie's decision and told the prime minister bluntly that Hughes was not wanted at the front. Major-General Steele was then sent home for his indiscretion in circumventing Perley and going straight to Borden. "Fighting Mac," Currie's successor, would not disappoint his Corps Commander from going to bat for him.[68]

When Macdonell took over as GOC of the "Old Red Patch" from Currie, he did not lose a stride in his approach to command and leadership style. Desmond Morton has styled him "a single-minded cheerleader for 'The Old Red Patch'" and "one of the few First World War generals to establish his eccentric personality among his admiring and occasionally embarrassed subordinates." Lieutenant James Pedley of the 4th CEF Bn had vivid memories of Macdonell telling officer cadets that he would "jump down the throats with spurs on and gallop the guts out of any officer" who failed to measure up to "Red Patch" standards. At gatherings, he would call out: "Who are you?" "The Red Patch!" "Are you with me?" "Yes!" – repeating the questions until he heard the right pitch of enthusiasm.[69]

After one year in command of "The Old Red Patch," Macdonell, a stickler for good officership, was not happy with the calibre of his junior officers, many being commissioned from the ranks within the Corps. "I wish every step taken to elevate the tone and status of Officers," he wrote. "Manners must be improved and character developed, in fact he must be a leader and an example in every way and inculcate the spirit of discipline by precept and example, creating an atmosphere by his own soldierly bearing and method of doing his duty."[70]

Leading from the front had always been his watchword from the first day he had commanded troops in action in South Africa. He successfully led the 1st Canadian Division from operations at Hill 70 and Passchendaele in 1917 through to the Corp's final spearhead to victory during "The Last 100 Days." He proudly rode alongside General Arthur Currie leading the Canadian Corps across the Rhine in December 1918, his beloved division first in the order of march followed by the Second Division. After a short occupation of three months, Macdonell would proceed to England with his Division in Spring 1919 and watch with tears as "The Old Red Patch" embarked from Southampton for Canada. His final orders to them read:

> I cannot view the breaking up of my beautiful 1st Division, the men of the Old Red Patch, with equanimity. It breaks me up too! I shall soon only be a memory to you. It will, however, I trust, be a pleasant memory of a Canadian General who believed in you, trusted you, gloried in your steadfast courage, discipline and truly wonderful achievements and who hopes none of you will be the worse for being Macdonell's Men.[71]

After the war, Macdonell served as the Commandant of RMC from 1919 to 1925, the first ex-cadet from the Army to hold that office. The College, which had been inactive during the war, needed a massive facelift and kickstart to get up and running again. The dynamic "Fighting Mac" with his strong views on what made a good officer was the man for the job. Macdonell went to work like a dervish, his first tenure of command quickly followed by an unprecedented second including a one-year extension. He supervised a revised and updated curriculum taking into account the hard-won experience of the Great War, expanding it from a three to four year program. He proposed and oversaw landscaping of the RMC grounds and new construction projects: the Education Building (West Wing); the memorial Sir Arthur Currie Hall; the Memorial Arch; the Holt skating rink; a Staff and Sergeants Mess; and, the renovation of the Fort Frederick Martello tower and creation of a Museum. He would also play an active role in securing RMC scarlet tunics to replace WW1 khaki, obtaining an official coat of arms for the College, and re-establishing the RMC *Review*. Macdonell would also serve at different periods of time as the Honourary Colonel of four different regiments: Lord Strathcona's Horse (Royal Canadians); The British Columbia Hussars; The

Stormont, Dundas, and Glengarry Highlanders; and, the Queen's Own Cameron Highlanders of Canada.

A tribute to "Fighting Mac" on his death in Kingston, Ontario at the age of 77 in the RMC *Review* he helped establish captured the character of this most unique Canadian general most aptly:

> General Macdonell was first and last a soldier. Martial in his air, his form and movement, witty, courteous, liberal, free of spirit. He had not the aloofness of a Kitchener; he was urbane, generous, hot-tempered at times, but full of the milk of human kindness. A stern disciplinarian yet he appreciated independence of thought in others, and fair play was second nature to him...He displayed, at all times, a loyalty which seemed almost excessive to his old College, the "Old Red Patch" and the Mounted Police, and an enormous pride in his Scottish ancestry, but after all, these were lovable traits in his character. He was a born leader and could always get the best out of his men; he knew exactly when to praise them for their work or when to condone their lesser shortcomings. When the occasion arose however, he could absolutely blast a man with the force of his invective, and yet, whether his tirade was long or short, he never tried to bolster it up with blasphemy or indecency...His friends were legion, won by his warm-heartedness, his manifest honesty and his courage. Also one who came in contact with him could not help falling under the spell of his enthusiasm and geniality or help realizing that they were in the presence of a great and gallant man. He never spared himself but was ever ready for the fray; he will always be counted among our great generals.

"Farewell, honest soldier!" [72]

---

1 G.G.D. Kilpatrick, *Odds & Ends from a Regimental Diary*, (Montreal: [privately printed] n.d.), p.24.

2 National Archives of Canada (NAC), RG 41, CBC Radio Transcripts of *Flanders Fields* (hereafter CBC), PPCLI, G.R. Stevens.

3 NAC, MG 30 E 241, *LCol D.E. McIntyre War Diary*, p.130.

4 Will R. Bird, *Ghosts Have Warm Hands* (Toronto: Clarke Irwin & Co., 1968) p.56.

5 G.R. Stevens, *A City Goes to War* (Brampton, ON: Charters Publishing Co., 1964) p.33.

6 A.C. Macdonell was born in Windsor, Ontario, 6 October 1864. His father was Samuel Macdonell, QC, LLD, DCL, first mayor of Windsor and Lieutenant-Colonel Commanding the 2nd Essex Battalion. His paternal grandfather was Colonel the Honourable Alexander Macdonell of Collachie, Invernesshire, Scotland who fought with the Royal Highland Emigrants (84th Foot) during the American Revolution, and later served during the War of 1812 as Asst Paymaster General. His great-grandfather also served in the Royal Highland Emigrants and Butler's Rangers. His family were Macdonells of Glengarry, a cadet branch of Clan Donald tracing their ancestry back to the Lords of the Isles and the great Somerled, who drove the Vikings out of Scotland.

7 Quoted in R.G.C. Smith, ed., *As You Were! Ex-Cadets Remember.* Vol. I (Kingston: The RMC Club of Canada, 1984), pp.115-121.

8 Ibid., pp.115-116.

9 Ibid., pp.115-121; see also unpublished *Biography of Sir A.C. Macdonell* by A.E. Kennedy-Carefoot in NAC, MG 30 E 20.

10 McGill University Archives, *Wallis Scrapbooks Collection*, Vol.1, MG 2039 (Acc. NO. 2627), Vol. I, Macdonell Letter to nephew Hugh Wallis dated 5 August 1914.

11 NAC, MG 30 E 20, *Macdonell Pocket Diary*, 1915.

12 Major-General A.C. Macdonell's undelivered Farewell Address to 7 CIB, 9 June 1917. *Wallis Scrapbooks Collection*. All letters cited from Wallis Scrapbooks Collection are addressed to his mother unless cited otherwise. Undated transcript of. Various portions of Macdonell's Address can be found in the various brigade unit's regimental histories and another copy is in NAC, MG 30 E 20. *A.C. Macdonell Papers*, annotated in the general's own hand, "This address was not delivered. I simply couldn't. So galloped off." The text reads: "The Brigade has always been a source of pride and joy to me, and when sorrow came into my life, and my only son was killed in action, it was the Brigade that saved me – pulled me through and comforted me... When we first mobilized at Mont de Cats, my job was to weld four of the finest battalions Canada sent to the war into a real Brigade – a military Machine that would run smoothly and well. Above all, I was anxious that the Brigade be a military family, we would all love and be proud of."

13 G.W.F. Nicholson, *Official History of the Canadian Army in the First World War: Canadian Expeditionary Force, 1914-1919*, p.134; Ralph Hodder-Williams, *Princess Patricia's Canadian Light Infantry, 1914-1919*, Vol I (London: Hodder and Stoughton, 1923), pp.91-92; R.C. Fetherstonagh, *The Royal Canadian Regiment, 1883-1933* (Montreal: The Royal Canadian Regiment, 1936), p.218; C.B. Topp, *The 42nd Battalion (CEF) Royal Highlanders of Canada in the Great War*, (Montreal: Gazette Printing Co. [1931]) 20; Stevens, pp.32-33.

14 Hodder-Williams, p.94.

15 NAC, MG 30 E 149 *Agar Adamson Papers*, letter to wife dated 5 January 1916. Unless otherwise stated, all letters are addressed to his wife.

16 Black Watch Archives, *Royal Ewing Letter Collection*, p.28. Ewing was the Lieutenant-Colonel commanding the 42nd by the end of the war and his letters covering the four years he served at the front, as platoon commander, Adjutant, coy commander, battalion 2IC and finally CO, are a valuable insight into this Montreal battalion.

17 NAC, RG 9 III D 3,Vol. 4938, *42nd War Diary*, 22 Feb 1916.

18 *Agar Adamson Papers*, letter dated 20 Jan 1916.

19 *NAC, MG 30 E 20,* Macdonell Pocket Diary 1915, 23 December 1915.

20 *Wallis Scrapbooks Collection*,Vol. I, Letter dated 6 Jan 1916.

21 *Wallis Scrapbooks Collection*,Vol. I, Letters dated 25 December 1915 and 28 June 1916.

22 B.A.H. Parritt, *The Intelligencers: The History of British Military Intelligence up to 1914*, (Ashford, Kent: Int Corps Assoc.,1983) p.1.

23 *Wallis Scrapbooks Collection*,Vol. I, Letter dated 30 May 1916.

24 As a matter of interest, Cecil Critchley's father, Oswald, and brother, Jack, were both Strathcona officers as well, his 66-year-old father signing up as the Regimental MGO at the outbreak of war with Colonel A.C. Macdonell's approbation. His younger brother Jack, a regular officer and senior subaltern served while Cecil himself was serving as Adjutant. Jack rose to the rank of lieutenant-colonel and command of the Strathconas, but was mortally wounded in March 1917. A second brother, Walter, served as a popular, hard-drinking major in Calgary's 10th Bn and later transferred to the RFC. Cecil would make a name for himself as an exemplary trainer as will be seen later and finish the war as a brigadier-general. A.C. Critchley, *Critch! The Memoirs of Brigadier-General A.C. Critchley* (London: Hutchison, 1961) in passim.

25 NAC, MG 30 E 20, *Macdonell Pocket Diary-1915*, entries for 23 and 28 December 1915.

26 NAC, MG 30 E 15, *Griesbach Papers*, File 11, Letter dated 12 May 1916.

27 Fetherstonagh, p. 225.

28 *Wallis Scrapbooks Collection*,Vol. I, Letters dated 9 and 29 October 1916.

29 *Wallis Scrapbooks Collection*,Vol. I, Letters to Harold Wallis dated 7 March 1916 and 16 April 1916 respectively.

30 William F. Stewart, "*Attack Doctrine in the Canadian Corps, 1916-1918,*" [unpublished MA thesis, University of New Brunswick, 1982), p.42.

31 D.J. Goodspeed, "Prelude to the Somme: Mount Sorrel, June 1916," in Michael Cross and Robert Bothwell, eds., *Policy By Other Means,* (Toronto: Clarke & Irwin, 1972) p.147.

32 Nicholson, pp.148-149.

33 Hodder-Williams,Vol. I, p.99.

34 Nicholson, p.150.

35 Stevens, p.44.

36 Ibid., pp.44-45.

37 Topp, p.52.

38 Hodder-Williams, pp.107-110.

39 NAC, MG 30 E 300, *Victor Odlum Papers*, Vol.21, Letter to John Nelson dated 4 Nov 1917.

40 *49th War Diary*, "Rpt on Counter-Attack Mt Sorrel" dated 8th June 1916.

41 Macdonell cited in Topp, p.58.

42 NAC, MG 30 E 20, A.E. Kennedy-Carefoot unpublished *Biography of Sir A.C. Macdonell*, p.157; McGill University Archives, *Col Hugh M. Urquhart Papers*, Acc. No. 393, Griesbach Interview.

43 Goodspeed, *op.cit.*, p.153.

44 Stevens, p.45; Hodder-Williams, p.133.

45 NAC, MG 30 E 20, *A.C. Macdonell Papers*, letter from Lieutenant-Colonel W.A. Griesbach dated 11 Feb 1917 and letter from Brigadier-General Griesbach dated 21 Feb 1919.

46 Ibid., A.E. Carefoot-Kennedy unpublished *Biography of Sir A.C. Macdonell*, p.155.

47 Nicholson, p.171.

48 Major-General Lipsett cited in Hodder-Williams, pp.165-166.

49 Topp, pp.77-78.

50 Critchley, *Critch!*, pp.68-69.

51 Ibid., p.69.

52 Ibid., in passim.

53 Nicholson, pp.247-249.

54 Hodder-Williams, pp.213-214.

55 Nicholson, pp.247-249; Hodder-Williams, pp.213-214.

56 Jeffery Williams, *Byng of Vimy, General & Governor General*, (Toronto: University of Toronto Press) p.149; for excellent discussions on the new German defensive systems and tactics see John A. English, *On Infantry*, (New York: Praeger, 1984) pp.14-17; Paddy Griffith, *Forward Into Battle*, (Ramsbury, Wilts: Crowood, 1990) pp.77-81; and Nicholson, pp.239-242.

57 Fetherstonagh, pp. 278-80; Hodder-Williams, pp.216-221; Topp, pp.123-124.

58 Black Watch Archives, *42nd War Diary*, "Summary of Operations" dated 13 April 1916.

59 NAC, RG 9 III, Vol. 2526, *1st Division WD*, Folder 52, File 7, "Rpt of 1st Division".

60 "Congratulatory Order on the Capture of Vimy Ridge" in G.G.D. Kilpatrick, *Odds & Ends from a Regimental Diary*, (Montreal: [privately printed] n.d.), p.26.

61 Stevens, pp.89-90.

62 Fetherstonagh, pp.287-292.

63 *Wallis Scrapbooks Collection*, Vol. I, Letter to Capt George Macdonald, PPCLI, dated 11 June 1916.

64 CBC, 49th Bn, A.E. Fowlie.

65 Stevens, p.90.

66 RCR War Diary, 10 June 1917; *42nd War Diary*, "Summary of Casualties, June 1916"; *49th War Diary*, 11 June 1917.

67 Stevens, p.91; see also Nicholson's comments in Nicholson, p.281, who terms the body count "excessive."

68 S.J. Harris, *Canadian Brass: The Making of a Professional Army, 1860-1939*, (Toronto: University of Toronto Press), p.126.

69 Desmond Morton, *When Your Numbers Up: The Canadian Soldier in the First World War* (Toronto: 1993), p.116.

70 NAC, RG 9 III, v.4024, f.8, Macdonell to 1st Canadian Division, 27 July 1918.

71 Quoted in Wallace G. Breck, "The Careers of Lt. Gen. Sir Archibald Cameron Macdonell, KCB, CMG., DSO, LLD, *Historic Kingston*, Vol 45 (1997), p.107.

72 Ibid., p.116.

CHAPTER 4

"A Brutal, Soul Destroying Business"

Brigadier-General F.O.W. Loomis and the Question of "Impersonal Generalship"

DAVID R. O'KEEFE

Major-General Fredrick Oscar Warren Loomis died in his bed on February 15th 1937, two weeks after his 67th birthday, victim of a heart ailment that reared its unprepossessing head in the years following the war. During the Great War, the contractor from the Eastern Townships of Quebec commanded the 13th Battalion (Royal Highlanders of Canada) during the bloody fighting of 1915, 2nd Canadian Brigade during two years of the most demanding, grueling and cruel combat to befall soldiers in western military history and finally, 3rd Canadian Division during latter half of the legendary "Hundred Days." Celebrated with the naming of a mountain peak in Alberta and immortalized in Edgar Bundy's *Landing of the First Canadian Division in St. Nazaire*, pallbearers of General's rank carried his flag-draped casket while mourners eulogized him at the time as a *great* Canadian General.

A few months prior to his death, the highly influential former military officer, historian and strategist, Major-General J.F.C Fuller produced a polemic 106-page monograph entitled *Generalship: Its Diseases and their Cure, A Study of the Personal Factor in Command*. In his treatise, Fuller addressed the decade-long groundswell of raw populist emotion that heaped sarcastic and caustic censure upon Allied generals and generalship alike. Characterized as "Butchers," "Bunglers" and "Jackasses" there is little, if any, understanding of the challenges faced by high command during the First World War. To some, the Generals formed a "faceless, uniformed pack," who unthinkingly sent the young, the untrained and the innocent to the slaughter with an indifferent wave of a hand. Their lack of intellectual curiosity, self-serving disposition, dispassionate, and wooden posture was of course a direct result of their station in life, their public school education, and insufficient grey matter needed to reconcile Victorian tactics and methods to an immerging modern world.[1]

Cutting through the vitriol, Fuller encapsulated both the indictment and alibi of a generation: "War with impersonal leadership," he wrote, "is a brutal soul-destroying business" and its impersonal nature is the root cause of poor generalship.[2] Central to Fuller's argument is his contention that Great War generalship suffered from a failure to understand the human condition in war that in turn translated into a fundamental disconnect between leaders and the reality experienced by the men under command. This "Chateaux Generalship" stripped the close contact of man and master, a contact that lifted the likes of Wolfe, Moore, and Wellington to the pinnacle of leadership success. The First World War, in Fuller's opinion, proved a watershed that plummeted military leadership into a moral malaise that in turn reduced the general to the likes of "managing director" or "prompter in the wings of the stage of war."[3] It was the caustic indifference towards the context of the war, and a world in transition, that drew Fuller's ire. "Neither a nation or an army" he wrote, "is a mechanical contrivance, but a living thing, built of flesh and blood and not iron and steel" and the failure to recognize, comprehend, and adjust to a new war reduced the "Chateaux Generals" to nothing more than "Ghosts who could terrify but who seldom materialized."[4] For Fuller, the pillars of "good generalship" reside firmly within a mind-set that favoured the creative over the conformist; a mind that possessed the astuteness to contextualize the experience of both early 20th Century industrialized warfare and a world caught in turbulent transition. In Fuller's opinion, the art of generalship was much more than simply "drilling troops"; it required courage of a physical, intellectual, and moral nature, and the ability to prepare and think ahead; to "peep around corners" and "spy out the soul of one's enemy." To regain the level of "True Master," a general needed to be "of flesh and blood, one who could see, who could hear, who could watch, who could feel, who could swear and curse, praise and acclaim, and above all who risked his life with his men, not merely issued orders mechanically from some well hidden headquarters miles and miles to the rear."[5]

With the recent re-discovery on his long-forgotten battle diary, Fuller's treatise serves as a provocative litmus test for a long-overdue re-assessment of this aspect of

Canadian generalship during the First World War and Loomis' generalship in particular. With this in mind, the following chapter focuses on his command of the 2nd Brigade in 1917 during the "year of professionalism" of the Canadian Corps – a year in which his Brigade played key roles in the tumultuous battles of Vimy Ridge, Arleux, Hill 70, and Passchendaele.

INSIGHT INTO THE HUMAN CONDITION

"Generals in these days of modern war and scientific slaughter lie on no bed of roses; the slightest mistake or miscalculation may mean the blood and lives of thousands of good men." This perceptive passage, recorded by Loomis following the August 1917 battle for Hill 70, was prompted by the noticeably worn appearance of Field Marshal Douglas Haig during a post-battle inspection of 2nd Brigade. To Loomis, his Commander-in-Chief's demeanour betrayed "the heavy responsibilities...on his shoulders" and that Haig "look(ed) like a man who realize(d) the full weight of his responsibilities...(and)...has the courage to meet them squarely in the face and overcome them."[6] Known for his mercurial temper, stern character, and hard-driving but competent generalship, Loomis never publicly removed his "mask of command" during or after the war.[7] What is immediately apparent from various passages in his battle diary however is that Loomis fully understood that virtue and frailty of the human condition form the main constant in war. In fact, Loomis' ability to contextualize the experience of the common soldier and civilian in relation to the responsibilities of command is one of his strongest attributes as a leader. From this insight into the human condition, everything flowed; and the results witnessed the creation of a confident, cohesive, aggressive and above all, successful 2nd Brigade.

Evidenced by the passages dealing with his arrival in the Lens sector in the summer of 1917, Loomis was clearly awake to the impact modern industrialized total war upon the human condition. "When an attack on any large scale is launched in this vicinity, unless the civilians are removed to the rear, there will be a holocaust indeed. Even during the comparatively small show that we are preparing, it is highly to be expected that these towns will be no harbours of safety."[8] Delving further in to their fate, he recorded:

> ...the peoples' terror is increased by the spasmodic, uncertain way in which the Germans throw over...small calibre shells, interrupted with heavies at odd times during the day and night at uncertain intervals, and one can well understand the mental anguish of Mothers whose children are out playing on the streets when she first hears the screech of a coming shell.[9]

It was in Lieven, a suburb of Lens however, where the full enormity of modern warfare struck home:

> The town...seen by day, presents a lugubrious but striking picture of the terrible effectiveness of modern artillery fire and of the devastation and destruction that the war has carried into the homes of our French Allies...the spectacle is heart-rending. What were once streets, teeming with industrious miners and tradesman, between rows of flourishing shops and peaceful homes, are now nothing more than interminable vistas of shell-pocked lanes bounded by repulsive looking mounds of wreckage and debris, broken bricks and masonry, rubble, plaster and the refuse from the inhabitants of the cellars beneath them... Splintered cradles and toys, the rags and tatters that once clothed the apple of some poor Mother's eye; household utensils, domestic and office furnishings; works of art and industry; everything in fact that one would expect to find in a large and flourishing civilized mining town, may be seen scattered broadcast and mutilated beyond repair amidst heaps of disintegrated masonry and litter of tin cans and empty bottles...The scene as one roams from house to house on a quiet day to view the wreckage is truly indescribable, and involuntarily fills one's mind with vivid pictures of the horrors experiences by the inhabitants of the town and wonders at their fate.[10]

Clearly, not a one-dimensional automaton, the build-up for the Battle of Hill 70 invoked a mixture of pride and foreboding that engulfed Loomis and betrayed his oft-concealed empathy for the common soldier and the fate of man. With the stage set and "all actors in their appointed places," Loomis understood that the latest "war dramas of the world...would bring many men face to face with the most critical moment of their lives, tested as never before." Soberly he concluded that, "to some will come death, a glorious death despite its sordid disguise of mud and slime; to some suffering and agony in many forms..." Likewise, the inhumanity of combat resonated with Loomis, and his battle diary is dotted with poignant passages of restrained emotion. In one instance, Loomis watched in awe as German and Allied aircraft duelled in the skies above only to witness one Canadian plane shot down in flames "like a wounded bird."[11] The violent descent ejected one pilot who plummeted to his death while the other remained trapped in the burning wreckage that crashed 150 feet from the Brigadier. "My orderly...carried in the body of the officer...who had fallen out earlier," he wrote, "he was horribly smashed and burned poor chap...I have since learned his name was Lt. John May."[12] In another case, when a German shell ignited a stack of ammunition near his headquarters that killed and wounded 40 men, he noted that, "some of the killed...were horribly charred and mutilated; their clothes burned completely off, as was their hair and skin, etc."[13] During a tour of the frontline trenches during the height of the battle, one incident drove home to Loomis his own sense of mortality when several shells burst within a quite a short distance, killing a private "who was standing so close...that I carried the stains of his blood home on my tunic."[14] Confronted by the horrors of modern war yet again later that day, Loomis recorded a mixture of shock and pathos as he witnessed four lines of German infantry steadily advance across No Man's Land under intense machine gun, rifle, and artillery fire. "The slaughter was something awful," he observed, "in a few seconds the few remaining Germans could be seen scurrying and crawling back into their own lines again, but very few could have reached safety."[15] Later that fall, just prior to the 2nd Brigade attack at Passchendaele, both the burden of command and nature of the fighting weighed heavily on his conscience:

> A survey of the ground recently captured from the Germans shows the various stages of our attack very clearly, and gives one a very vivid impression of the terrible character of modern warfare. As we approach the area captured in the first advance, we see aeroplane wrecks in various stages of demolition; some almost intact, where the wounded pilot had been able to make a good landing, others crumpled up through a fall and still others just a jumble of twisted iron and charred wires and canvass. Wrecked tanks freely perforated with shell holes and more than half

> buried in the mud mark the course followed by these monsters on their work of rooting the enemy out of his pillboxes and strong points. Up to this stage the battlefield had been fairly well cleared, but further forward where, until we captured the ridge, the ground was under constant observation could not be effected, and we come upon more gruesome signs of battle. Dead bodies, friend and foe, lying in shell holes and partially sunk by their own weight into the mud, show the earlier stages of the attack while later stages can be recognized from the fresher appearance of the corpses and the blood stains that still mark the ground and colour the water where they made the supreme sacrifice for their country. Up towards the front line are the still warm bodies of those who fell a few days ago and who paid for the Canadian victory of Passchendaele with their lives. It is to be hoped that after we have cleared the enemy from the last high ground which we are about to attack that these heroic remains may be collected and given the decent burial that is the right of every soldier killed in action.

This well-balanced approach and empathetic nature extended to the life of the soldier in the front line and the staff as well. On many occasions while making regular visits to the front lines, Loomis remarked upon the peculiarities of soldiering and life in the trenches. In December 1916 near the inundated town of Brouay, the Brigadier was "shocked and surprised" to see that "the boys absolutely refused to allow their spirits to be dampened...and it did one's heart good to see the excellent manner in which they accepted very trying conditions."[16] A little under a year later in the morass of Passchendaele, he noted that the rain made conditions "most heart-breaking" and added "immeasurably to the difficulty of the already hard task facing the men." In addition to the weather, the "continual quietness" of the enemy before the Canadian attack caught his imagination as well:

> Rather a peculiarity in the soldier's character is to be deduced from this, and is attributable no doubt to the fact that he is always under constant strain: if things are lively, he grumbles because the enemy will not leave us alone; if things are quiet, he is worrying as to what the enemy is contemplating putting off on us: in fact, I think, that our men are in a more happy frame of mind when there is a little shelling going on (not

> too close) as then he is not afraid of anything worse, but when things are quiet his imagination gets working and this war has produced such devilish methods of killing and torturing people that it is pardonable for them to wonder if the Germans are hatching some new fiendishness.

Despite his sensitivity, the mask of command remained in place, and Loomis balanced his emotions and compassion with the responsibilities and demands of command and leadership. On at least two occasions during his tenure, soldiers in 2nd Brigade faced court martial proceedings that produced a death sentence. In one case, Loomis was relieved to report that despite the verdict, extenuating circumstances allowed the commutation of the sentence to imprisonment.[17] In the other however, the Brigadier refused to bend as the accused had received a reduced sentence for earlier breaches of conduct in this regard. "It was a particularly bad case," he wrote, "there were no extenuating circumstances" and "Private H Kerr of the 7th Battalion, suffered the extreme penalty."[18] Taking no comfort in this decision, Loomis consoled himself as "fortunately, the (is) first time that this proceeding has been necessary in my Brigade."[19]

At times however, he expressed his frustration with his failure to conquer the various quirks and foibles of the human condition such as the "superstitious dread" of a Friday the 13th or more importantly, the failure of some to adhere strictly to orders that as a result, caused needless casualties.[20] With a mixture of pathos, disgust, and paternalism, Loomis commented on the failure of some soldiers to dig-in after their attack on Hill 70. "Rather than work they stayed in the German trench which the enemy subsequently shelled very heavily causing very high casualties to the occupants while a platoon, which had dug in as they were told to and had used their shovels freely, escaped with practically no casualties at all." Later, tempering this assessment, he admitted that "it is one of the hardest things to get a soldier to work much after the excitement of the attack has spent itself and he begins to relax; though it is one of the most important, and all soldiers should be made to realize it for their own personal safety as well as for that of the positions they have paid so much to capture."

As the preceding passages reveal, Loomis possessed both an understanding and fascination with the effects of war on the human condition and its practical application to the successful prosecution of the war. They key however to his success in leadership was the ability to weight that at times, conflicting nature of command and human condition.

TRAINING AND PREPARATION FOR BATTLE

Regarded as one of the bloodiest and most futile exercises in modern history, the campaign on the Somme acted as a catalyst for a more professional Canadian Army that both engulfed and engrossed Loomis. The results of the bitter lessons produced a higher level of competence through training and reorganization as well as the application of new or developing technology and tactics, something Loomis' receptive character and intellectually curious disposition proved perfectly suited.[21] Fully aware of the new challenges of leadership, the hallmark of Loomis' generalship lay in his ability to prepare and organize for battle. One glance at his operational orders, war diaries, or battle diary, and it is clear that he possessed an almost legendary penchant for organization and detail that laid the cornerstone for success in battle. Cut from the same cloth as Arthur Currie who believed that if training was right, leadership was what it should be, and preparations complete, the men could do anything.[22] In response to a query about the lessons of the Somme, the Brigadier proved unequivocal in his assessment: "the answer...is TRAINING, a training that will not only result in securing discipline and higher technical knowledge, but will also secure great resourcefulness in Officers, NCO and men."[23]

The evolutionary nature of infantry warfare within the context of the early 20th Century was not lost on Loomis as his training plans clearly indicate. Designed to develop "the individual's initiative as well as his muscles," training ensured that the men were "thoroughly up-to-date with all the little new wrinkles which are constantly coming out" and addressed the essentials of the efficient fighter – something he termed the "versatile soldier."[24] Loomis firmly believed that new training freed "the minds of all Regimental Officers of the old ideas and obsessions of trench warfare" and observed that units did not worry any longer about their flanks, but instead pressed the advance.[25] The consequence was the enthusiastic acceptance and promotion of the concepts of flexibility, initiative, and aggressiveness – something not seen in the days of Kitchener's Army but facilitated in earnest by Loomis and others within the Canadian Corps.[26] As early as August of 1916, Loomis's training syllabus included advanced skills such "bomb throwing," "machine-gunnery," "signalling," "scouting and observation," as well as the regular training in "musketry," "bayonet exercise," and "skirmishing."[27] Later that summer, Loomis pitted his battalions against each other in a series of exercises involving both the attack and defence in conditions resembling "those which would obtain in actual warfare."[28] The great lesson of the Somme for Loomis however, was the need for clever and efficient use of direct and indirect firepower and he immediately implemented numerous exercises "illustrating the principles of attack...employment of artillery and machine-guns, communications, use of the ground...fire control...and flanking fire."[29] Other training sessions stressed the speedy writing and issue of appreciations, orders and reports, as well as close all-arms cooperation and reconnaissance, and warned of the dangers of stereotyped practices and of over-tasking sub-units.[30] Each Tactical Exercise Without Troops (TEWT) concluded with a wrap-up by Loomis to ensure that all understood the training with no key issues missed or false lessons learned.[31]

In the New Year, 2nd Brigade undertook a "vigorous" resumption of training with the view of making 1917 the "premier year in every way in the history of the unit."[32] By this point, the Canadian model of training and preparation reached the operational level of war with the development of an attack doctrine that incorporated technological, tactical, and organizational changes.[33] Aware of the subordination of "battle command" to "battle preparation", Loomis agreed with Brigadier Victor Odlum that "Our fights are won or lost before we go into them" and was quite in line with General Julian Byng's mantra that "training in peacetime is the most important part of soldiering...In wartime it is only second to operations, and operations can only be as good as the training is efficient."[34] In response to the new system, Loomis believed that the efficiency and striking force of his battalions could only be maintained during the severe fighting to come by "most careful and sound organization and by a high standard of training and discipline."[35] To illustrate this attention to detail it is worth focusing on one aspect of his preparation for the Vimy offensive in the spring of 1917. Not happy with the maps produced by Division, Loomis complained bitterly that they proved "entirely inadequate to provide the understanding and instruction necessary...for an operation of this magnitude." In response, he took the initiative and put his draftsmen to work at once on new maps that included the work accomplished and that currently underway. The results were comprehensive and impressive: German emplacements, observation posts, stores, dumps, dug outs, trenches, and gaps in wire, locations, and character of trenches, gun positions, areas covered by machinegun and trench mortar fire, all appeared. On the Canadian side, the list was exhaustive: assembly positions for Canadian infantry, jumping off trenches, contours of the ground, objectives, consolidation positions, and boundaries of the brigade. In addition, main communications trenches, ammunition and Engineer supply dumps, communication trenches, tunnels, mule trails, tramway routes, dressing stations, headquarters, emplacements, hot food depots, deep dug-outs, observation posts and fire trenches were all compiled to "be of the greatest possible use when needed."[36]

A stickler for details, Loomis proved a strict taskmaster as well. On one of many inspections of frontline trenches before Vimy, the "lack of interest" shown by some officers and men to "work of improvement and tactical importance" distressed him. "Few," he wrote, "seem to realize that these 'working parties' are for the benefit of the men themselves...(and)...the necessary construction put forward, being designed to save lives and to make defensive and offensive operations more effective."[37] Three days prior to the main assault on April 9th, he was again "disappointed" that little maintenance had been done on communication trenches: "Although I much regretted to do so," he confided in his battle diary, "I called on large working parties in order to restore the trenches before the attack." For Loomis, not simple success prompted this action but rather "work of this kind means men's lives, and of course its accomplishment is vital."[38] Eventually as the operation approached, Loomis concluded with an air of satisfaction that:

> Great attention has been paid to the organization of our units, the ensuring of plentiful and easily accessible food and water and ammunition supplies, the provision of adequate engineering material for the consolidation of captured positions, the rapid evacuation of sick and wounded, the establishment of Hot Food depots at convenient positions, and many other features. Provision has been made for emergencies of almost every kind and there will be carrying parties standing to at all times awaiting instructions for the carrying of ammunitions and food to required points.

Loomis was not the only one satisfied. Unlike other Brigades that averaged 20 comments from Currie, 2nd Brigade drew only five. According to Lieutenant-Colonel Kenneth Radley, who penned the definitive history of the 1st Canadian Division, his plan was "just plain well done" and contributed in due course to the stunning victory at Vimy.[39] Success followed again in mid-August when the Canadian Corps captured and held Hill 70, dispatching five German divisions in the process. To Loomis, this victory confirmed and reinforced the vital lesson of Vimy: "prepare thoroughly and in detail, train hard and build in redundant control and communications. Consolidation was a key factor. Had the attack not been so well planned and resourced, which enabled rapid consolidation; determined counter-attacks might have achieved some success."[40] According to Radley, Hill 70 was "a model of sound tactics and superb planning" and Loomis' preparation and planning proved well thought-out, comprehensive, organized, logically presented, and maintained a clarity that set the tone for the operation itself.[41] Crisp and clear with no ambiguities, it emphasized speed and a tight decision-making cycle; not only were his battalions to get inside the German barrage but the German decision-making cycle as well. His four-wave assault was on one hand methodically planned to keep the troops and the tempo of the battle under his control, but flexible enough to ensure initiative at the battalion, company and platoon levels. Again, counting on the latest in tactics and training Loomis urged his troops not to stick, but to use manoeuvre to overcome or circumvent obstacles using weapons and resources at their disposal. In what would become standard throughout the Canadian Corps and proved so successful at Vimy, Loomis ordered an exact replica of the German trenches and defences constructed from aerial photographs. Loomis could not contain his excitement at the idea of the men acquainting themselves with "every detail" of the ground to be covered: "Our men will know where to look for Machine Guns, Trench Mortars, and Infantry Dug-outs etc, and what trenches to block. In fact, it gives the men a tremendous confidence to think that they are attacking over familiar ground and it raises the probabilities of success a hundredfold."[42] In addition, Loomis dispatched a staff officer to the front lines to fix natural or structural features on the lines of the upcoming advance to serve as guides for the men going over the top. "One of the greatest dangers in an attack in modern warfare where the ordinary structural and vegetable features are all leveled or undergo vast changes in a single bombardment," he wrote, "is that of losing the sense of direction, and so making gaps in the line and causing dangerous crowding in the attacking ranks."[43]

In his Special Instructions, Loomis stressed the importance of receiving information in a timely fashion, clearly demonstrating his awareness of the need to reduce the Fog of War to a mere mist.[44] When the Battalions settled in on the captured positions, Loomis insisted that they remain in touch at all times, forwarding clear and concise reports with attached maps whenever possible. To keep a close eye on developments before and during the assault, the Brigade laid 700 yards of cable and constructed various Observation Posts, report centres and an advanced headquarters with a bevy of runners so Loomis would never lose personal touch with the situation at the front. When these failed to produce adequate results at the most crucial point in the battle, he ventured forward and conducted a personal reconnaissance of the new positions in an act of bravery and generalship that garnered public praise from the divisional commander. Once on sight, he counted on preparation and innovation to tackle the more intricate problems of mopping up, carrying parties, and stretcher-bearers that spelled disaster during the battle of Mount Sorrel. Loomis' concern for the Hill 70 operation was that the large area assigned to 2nd Brigade required the employment of almost all the assault troops to ensure a proper supply of material. "Mopping up" in Loomis's opinion, was a "serious question," and with the enemy's use of new tactics such as "defended shell-holes" and "isolated strong points" he decided the best way to handle this problem was to overlap his assault forces ensuring that the men would complete the work after the capture of their objectives. His foresight and innovative approach worked well under a most severe test; both Loomis and his Battalion Commanders were satisfied with this solution as it permitted the relief of his lead battalions – who had suffered heavy casualties – and allowed them to retain cohesion and resume their attack promptly at zero hour.[45]

Likewise, training and preparation for the Passchendaele campaign was "state of the art." Taking into consideration the nature of the terrain and the expected German defences, specialized work such as attacking pillboxes, fortified houses, and strong points, wire cutting, and anti-gas training, received special attention. Based on a realistic foundation, Loomis fused the nature of the terrain, the fighting and desired objectives, with the form of the German defence to create a paradigm for success.[46]

Cohesion was the constant thread. To ensure this, his staff created a plasticine model of the assault area that each officer and NCO in the Brigade visited before the attack covering such topics as boundaries, objectives, scheme of attack, the nature of the ground, and likely German reactions.[47] Daily, officers reconnoitred the situation and the ground leading up to the push on November 10th and Loomis presented his Battalion Officers in Command (OC) with timely information of the battle unfolding up ahead and addressed such matters as formation for attack, arrangements for mopping up and communication, equipment to be carried, carrying parties, barrage time table, consolidation and relief, leaving fuller particulars as to boundaries, frontage and depth of attack to Division.[48] The time afforded for training and preparation allowed the OCs of each assault battalion to familiarize themselves with terrain and Loomis' plan as well as the overall form and tone requested by all levels of command. The troops too had over a full day to adjust to the ground and make final preparations. Indeed, all found themselves on the same page and success soon followed.[49]

For Loomis, this detailed preparation enhanced battle procedure, command, and control; two elements that proved particularly decisive in the prevailing conditions of mud and blood. When 2nd Brigade's turn came on November 10th, Loomis' two-battalion assault secured its objective by noon, just six hours after Zero hour. No infantry counter-attack ensued; instead, a fierce and protracted artillery barrage shook 2nd Brigade but did not eject them from their positions at Vindictive Crossroads and Venture Farm.[50] In Loomis's assessment, the two outstanding features of the Operation were the continuity and intensity of the enemy's artillery fire and the "demoralizing effect" of the retirement of the unit protecting their left flank. In his official report, Loomis wrote that fact that his brigade withstood intense artillery and machine gun fire and held their objectives in spite of the demoralizing effect of the withdrawal...appear(s) to be sufficient proof of sound training and methods."[51] Major-General MacDonnell, the Division Commander, concurred with this assessment and added that success lay with the initiative given to each Brigade to study its own problem, work out the best plan, and allow everyone associated plenty of time to study it and know their place in the chain of command.[52] If indeed, as Ian McCulloch suggests, competent Brigade commanders are "cool, calm, collected" and work "behind the scenes to ensure a coordinated effort...to make sure his troops (are) in the right place at the right time with the right resources," it was clear during the course of 1917 that Loomis succeeded in this regard.

THE INNOVATIVE MIND

By 1917, Currie, taking his cue from Byng, created a positive and receptive atmosphere in the Canadian Corps that fostered the free flow of ideas from below and inculcated the same spirit in his subordinate commanders. Loomis, possessing an inquisitive intellect, proved to be one of the most innovative of Canadian Brigadiers.[53] One lesson Loomis learned from the Somme experience was that new technologies and methods put tactical command and control in the subaltern's hands and he constantly searched for ways to harness new technology and tactics in an effort to improve the performance of his troops and counter ever-evolving German tactics.[54]

Loomis was not a foggy intellectual or technocrat enamoured with technology for its own sake; his utilitarian and practical nature ensured that he did not get lost in the details. It is clear that he understood that a symbiotic relationship existed between the reality of his troops in the frontlines and the technological requirement for success. Principally, the Trench Mortar and Lewis Gun caught his attention due to their force enhancement capabilities and primary relationship with infantry tactics. In the fall of 1916, Loomis observed, "as far as it was possible to do under the prevailing conditions, the gap on our front was closed by means of small groups of men placed in independent shell holes and armed with Lewis Guns before the line was turned over."[55] Prior to the Hill 70 operation Loomis investigated the merits of new design for the Lewis Gun. Noting that the new invention by a Private from the 7th Battalion provided the "very great advantage... especially in firing from the sling while advancing," his great interested stemmed from the fact that it "removes from No.1 the duty of removing the empty magazine in action, so that he can keep his attention concentrated on this target."[56]

As for the Trench Mortar, the practical reason that they could "crank up enemy working parties, and...keep him busy; repairing the damage they continually caused him" proved the appealing factor.[57] Later at Clarency, Loomis harnessed the striking power of massed Stokes Guns to eliminate sniper activity."[58] In this case, Loomis instructed his scouts and observers, in consultation with his Intelligence

Officer, to "pay close attention" to the points from where the sniping emanated. When an enemy sniping post was accurately located, a "direct and obliterating stream of bombs" from the Stokes guns followed. "This," Loomis wrote with an air of satisfaction, "has been found to silence the enemy.[59]

Forever seeking to improve performance on the battlefield, the capture of Lieven with its reinforced cellars, buildings, fortified barricades, and gun pits strewn throughout the town offered Loomis an intellectual playground of "the greatest value" from a tactical, strategic, and technical perspective.

> With our first hand knowledge of both sides of the question we are excellently able to judge the effect of our artillery fire; to calculate the value of different kinds of protection against shell fire; to test the hitherto uncertain value of reinforced concrete as a shell resisting material; to determine the most efficient means of reinforcing cellars and houses and to establish the best guns, shells and fuses to use for their destruction. We have also, no doubt, picked up a lot of new wrinkles in the constructions of gun pits and observation posts from the enemy's many abandoned models and from the examples of this defensive ingenuity, which were left intact at the hurried retreat of his troops.[60]

Another example of his innovative disposition came during the run-up to the Hill 70 operation when his constant vigil on German working parties revealed "the enemy...availing themselves of intervals when there is no artillery, trench mortar, or machine gun activity on our front."[61] In response, Loomis drew up an artillery scheme that coordinated all weapons to cover the whole area with 24 hours of constant harassing fire. To Loomis, the reasons were self-evident:

> The enemy has been very actively engaged recently in repairing and reorganizing his defences where they had been damaged by our recent artillery preparations and the postponement of our operation has given him an opportunity for this, our fire having slackened, so that a lot of the work will have to be done over again. Under this new scheme, when we do any damage to his defences our coordinated fire will make it very costly in men for him to effect repairs and will greatly effect his morale, for when the artillery is not firing our Machine Guns Bde [Brigade] , or Div [Division], will be, and if none of these doing so, then the medium

> trench mortars or Stokes Guns will be covering the area; this will be kept up during the 24 hours of very day until the attack is launched and all governed by time-table.[62]

Division quickly approved his scheme without hesitation or reservation.

The advent of gas warfare and the impact of aviation on the battlefield made deep impressions upon Loomis. Cleary, the Brigadier understood firsthand the nature and impact of this unwieldy and hideous weapon having suffered through the first gas attack in history as OC of the 13th Battalion in April 1915. His return to the Ypres salient a year later evoked bitter memories of this most sinister weapon. "Canadians need no warning of the deadly quality of this weapon of modern warfare, as it was in this very salient that, just over a year ago, without protection of any kind they withstood and flung back the first attack launched the Germans behind a screen of gas."[63] His innovation in this area came in the form of "gas alerts" that became standard operating procedure within the brigade when weather conditions proved favourable to gas attack. *Gas alerts* were different from gas alarms that heralded an actual attack. The *gas alarm* sounded from the first point where gas appeared. Seconds later, the whole area sprung to life with the "clanging of gongs, rails, klaxon horns and every possible device for conveying the arm..." At this point, all ranks donned gas helmets and took standard gas precautions. However, if it were not for the *gas alert* preceding the *gas alarm* pre-cautions taken would be of little avail as Loomis calculated that a gas cloud moves so swiftly that very few second are available before the noxious fumes arrive; in all 9-20 seconds, according to the wind was the standard interval between discharge and arrival of the cloud. In this case, "unless helmets are handy" he wrote, "there are sure to be a large number of casualties."[64]

In the skies above, aviation, in particular aerial photographs, captured Loomis' tactical imagination.[65] During the planning phase of each major Brigade operation in 1917, Loomis relied heavily on aerial photographs without losing sight of the fact that nascent technology is a doubled-edge sword. "The temporary advantage which the new fast aeroplane, used by the Germans which is giving them the navigation of the air, has enabled them to spot a good many of our gun positions, dumps and routes of traffic, with the result that these have been shelled more or less."[66] In response, Loomis allowed a subordinate to experiment with a Stokes mortar in an anti-aircraft role where his one and only attempt proved somewhat successful.[67] By Passchendaele, Loomis clearly understood the emerging role of air power as a force enhancer in connection with ground operations – or in this case, the likelihood of enemy interdiction and reconnaissance in response to the Canadian push. To this he warned his Battalion Commanding Officers (CO) to expect German reconnaissance planes overhead and ordered them to turn "all available Lewis Guns and as much rifle fire as possible" to keep them away and "interfere with the using of their counter-attack battalion to the best advantage."[68] In preparation for this, Loomis arranged with the Royal Flying Corps to have aircraft fly within rifle range of the ground over the Infantry Battalions while they trained. "This gave each man," Loomis wrote "a good idea as to when he should, and when he should not, fire at aircraft."[69]

Ever the sharp critic of both technical and tactical innovations, Loomis was never shy to voice an opinion or share the results of experiments in a judicious fashion with those concerned. In the wake of the Somme fighting, Loomis attended a demonstration by 3rd Canadian Division "purporting to illustrate the consolidation of a trench after being demolished by artillery fire" but the results, according the Brigadier-General, were "disappointing." His dismay stemmed from a lack of professionalism on the part of the demonstrations organizers as "there was no effort made to get down deep; and the work was but indifferently carried out."[70] A little over a week later, he organized a surprise test for the Brigade artillery in which 10th Battalion would request a surprise hurricane bombardment of all available guns on specific sections of the enemy's trench system at three pre-arranged hours. The idea was to evaluate how well prepared the artillery were to cooperate in an emergency such as a sudden surprise call or SOS. "The first two shoots were far from satisfactory," he recorded "and the third...left much to be desired." His chagrin at the results stemmed from the fact that one battery shelled Canadian positions while others fired at the wrong targets altogether. Not impressed by the "carelessness in gun-laying," he recorded in understated fashion that "better liaison with the artillery and communications with the FOO [forward observation officer] were needed."[71]

Despite his mercurial temper, Loomis could also be discreet and diplomatic with his criticisms. In response to a Corps Intelligence Summary during the battle of Hill 70 which touted the barrage on August 17th as "very successful" Loomis reported that while overall this was correct, it had not been effective on the reverse slope of his second objective as fire had burst "about 100 feet too high" failing to neutralize the enemy.[72] These were not just empty complaints, or attempts to distance himself from the wealth of casualties suffered in the attack, (he sent the report to the operations staff and not the Division commander) but indeed constructive criticism that revealed his grip on the ever evolving Canadian attack doctrine. A prime example of which came while 2nd Brigade prepared for the move into the Passchendaele sector in the fall of 1917. This time, Loomis' target for criticism was the rolling artillery barrage that did not completely suit the infantry's needs during the Hill 70 Battle. Here, he noticed that the barrage proved "more efficient in the earlier and nearer stages of the fight" than it did on the more distant objectives. The result was dispersion of the barrage and gaps in covering fire that enabled enemy garrisons to swing into action and delay the advancing troops who needed to keep up with the barrage. His solution was twofold: first, gunners had to place more emphasis and concentration on distant targets during their preparation period and second, reduce the timetable and progress of the barrage to correspond to both the contours of the terrain and the pace of the infantry moving from shell-hole to shell-hole during the advance.[73]

According to Loomis, the "most important" lesson of the Hill 70 battle within the context of the new attack doctrine was the question of immediate relief of the assault troops and eventually the Brigade itself. Quick consolidation and preparation for defence of the newly won ground formed the lynchpin of the Corps plan to defeat the inevitable German counter-attacks. As such, Loomis focused his attention on the relief of the assault troops and understood that the nature of the fighting precluded the relief of the Brigade itself leaving it *within* the Brigade to solve.[74] "Troops who have had heavy fighting are," he wrote, "necessarily, reduced in numbers, in physical force, in ammunition and in water and supplies." "If the fighting is continuous," he added, "it is impossible to have carrying parties going back and forth successfully." His remedy called for troops specifically trained in the relief and defence of captured positions, to come up at the first available opportunity loaded with supplies, engineering materials, water and rations. These troops, in his estimation, "will work with vim, will quickly increase the defence, and... hold the position secure" resulting in fewer casualties than witnessed in the Hill 70 fighting.

At Passchendaele, Loomis exhibited a well thought-out approach to a tactical conundrum faced by the terrain and weather conditions. Here, it was of "vital importance" owing to the conformation of the ridge and subsidiary spurs, and to the fact that the ground between the spurs was impassable on account of mud and water, that the advance had to be made up the spurs over restricted ground well known to the enemy. In these killing grounds, Loomis warned that the Germans appeared to direct machinegun fire down the slopes through the barrage along the contours of the spur, which conform "almost perfectly" to the trajectory of the bullets. To remedy this, Loomis, considered the advisability of still further retarding the rate of advance in order to allow the assaulting troops to make their way forward slowly, taking advantage of the shell-holes for the duration of the advance, literally crawling from one shell-hole to another and keeping their bodies as flat on the ground as possible. To aid in the attack, Loomis ordered a reciprocating barrage to rake the forward area starting with a fall of six minutes on the line, two minutes 100 yards forward, then back for four minutes on the original line and then forward 100 yards permanently, the infantry then to go forward the 100 yards.

In the wake of the Passchendaele campaign, Loomis issued a 33-page report on the capture of Vindictive Crossroads and Venture Farm. The lessons drawn from the operation indicate an awareness and understanding of the evolving operational art. According to Loomis, success emanated from the "initiative" and "resource" shown by all commanders from platoon level upwards that resulted in the systematic approach to mopping-up isolated and bypassed German positions. In addition, he noted troops advancing by bounds and a slow rate of fire for the barrage (that allowed junior officers and NCOs the opportunity to selected cover, time their rushes and deal with choke points and other obstacles) as outstanding characteristics of the operation.[75] Part of the "overwhelming" success he recorded, stemmed from the special effort made to provide each individual taking part in the attack with a comprehensive appreciation of the general objectives and the role played by his battalion, company, platoon, and the individual himself.[76] In addition, each soldier was reminded of the need to keep close to the barrage, to use the prone position in dead ground and shell-holes when not rushing forward, to secure objectives quickly and select the best tactical position for consolidation where they could dig-in, hold their positions, and yield no ground.[77] According to Loomis, the numerous instances where Other Ranks took Command of platoons and NCOs of Companies and carried out the work intelligently, and with discretion and judgment, were examples of the "careful instruction" and the Currie-inspired taking of all personnel into confidence.[78]

CONCLUSION

Perhaps there is no better indication of Loomis' understanding of the human condition in relationship to generalship than is witnessed by the after-action report he penned following the bloody battle for Hill 70. As much as it is an astute record of events and lessons learned, Loomis exploited the forum to convey and reinforce the total team effort he desired within his brigade. Skillfully, the Brigadier infused a cohesive tone and reinforced the virtues of all-arms cooperation by singling out the efforts of both units and individuals. His underlying motive without a doubt was to provide a blueprint for success – a counter balance for his usual sharp criticism, pointed remarks and mercurial temper. First and foremost, he lauded Divisional and Corps artillery for "the preparation of the area for attack" and for their "most effective" moving barrage and for their "promptness in answering SOS calls throughout the operation. In addition, he cited No.1 Motor Machine Gun Brigade for their harassing fire during the preparatory stage while the 2nd Field Company Canadian Engineers (CE) who built communication trenches and lay of wire merited mention as well.

Once he dealt with the actions of the support units Loomis turned his attention to individuals under command. His brigade-major received the lion's share of the credit with Loomis' singling out his exhaustive inspection of the front line areas, and his penetration of No Man's Land in daylight in order to inspect a proposed jumping-off trench. Likewise, Loomis singled out his staff captain intelligence, for personal bravery and as well as his staff work which was "most commendable." In particular, the Brigadier pointed to his two "valuable reconnaissance" of the forward area after the attack, under very heavy shellfire" and the results of which proved "most useful." From here, Loomis turned to the work of those responsible for A and Q duties, stocking of dumps, laying of cables, gas-proofing, and continued with Liaison and Forward Observation Officers, the clearing of wounded, the building of strong points, the clearing of mines and booby traps, carrying parties, intelligence, map marking, sentry and guide duty, communications, the preparation and distribution of operational orders, and even those who prepared 6000 meals in the Brigade Food Kitchen. Loomis heaped the majority of his praise, however, upon the battalions; in particular their decision-making, courage, command and control, initiative, and forethought, as well as their ability to act and react accordingly in a given situation of extreme distress and maintain communications with his headquarters. To any observer Loomis' understanding and commiseration with their efforts, along with a well-deserved recognition and whole-hearted appreciation of teamwork and the ability to disseminate this to subordinates in a subtle and indirect fashion, provided a rock-solid blueprint for that paved the way for the success of 2nd Brigade in 1917.

---

1 Fuller, J.F.C. *Generalship: Its Diseases and Their Cure: A Study of the Personal Factor in Command.* (Harrisburgh, PA: Military Service Publishing, 1936).

2 Ibid.

3 Ibid.

4 Ibid.

5 Ibid.

6 General F.O.W. Loomis Battle Diary (LBD), 24 August 1917.

7 Canadian Letters and Images Project, Diary of Alfred Herbert John Andrews <www.canadianletters.ca> accessed May 1st 2008 "On Oct 24 we were out as usual when the Brigadier General Loomis rode up. He watched us for a bit and then asked me how I fired at aeroplanes. I told him I had asked that question of everyone I sent to a school and none of them had any idea and the only thing I could suggest was to set up a barrage in front of the plane and hope that it would fly into it. That I was sure it was only a matter of luck if a Lewis gun brought down a plane. He asked me how many anti aircraft sights I had and I said 3. He turned to his Brigade Machine gun officer and said, "Were not 4 issued?" I admitted that 4 were issued but only 3 were Lewis gun sights and the 4th was for Vickers. The Brigade officer insisted there was no difference.

I respectfully differed with him but the Brigadier said he preferred to accept the statement of his own officer and that if I ran a business the way I did my guns I'd be broke in a month. He gave me a general bawling out before the troops and then rode away."

8 LBD July 19th 1917.

9 Ibid.

10 LBD September 10th 1917.

11 LBD January 23rd 1917.

12 Ibid.

13 LBD August 9th 1917.

14 LBD August 16th 1917.

15 Ibid.

16 LBD December 11th 1916 .

17 LBD August 25th 1916.

18 LBD November 21st 1916 and Godefroy, A.B. For Freedom and Honour? The Story of the 25 Canadian Volunteers Executed in the Great War (Nepean, ON: CEF Books, 1998).

19 LBD November 21st 1916.

20 LBD July 13th 1917.

21 Library and Archives Canada (LAC). McCulloch, Ian. *The Fighting Seventh: The Evolution and Devolution of Tactical Command and Control in a Canadian Infantry Brigade of the Great War.* (RMC: 1997) p.144.

22 LAC Radley, Kenneth, *First Canadian Division, CEF 1914-1918: Ducimus* (*We Lead*). p.3.

23 LAC RG 9 III C3, Vol. 4051, Loomis to Division quoted in LAC Radley, p.246.

24 LBD January 27th 1917.

25 LAC Radley, p. 367.

26 LBD January 27th 1917.

27 LBD August 2nd 1916.

28 LBD August 24th 1916.

29 McCulloch, p.144.

30 LAC Radley, p.324.

31 Ibid.

32 LBD January 2nd 1917.

33 McCulloch, p.183.

34 Critchely, A.C., *Critch!: The Memoirs of a Brigadier General* (London: Hutchison, 1961).

35 LBD September 5th 1917.

36 LBD March 31st 1917.

37 LBD March 18th 1917.

38 LBD April 6th 1917.

39 LAC Radley, p.137.

40 LAC Radley, p.239.

41 Ibid.

42 LBD July 17th 1917.

43 LBD August 14th 1917.

44 LAC RG 9 III War Diary 2nd Infantry Brigade September 5th 1917 Operation Order 212 (Hill 70).

45 Ibid.

46 LBD October 26th 1917.

47 LBD November 7th and 9th 1917.

48 LBD November 7th 1917.

49 LAC Radley, p.241.

50 Ibid. p.242.

51 LAC RG9IIID3 Vol 4872 WD 2nd Canadian Brigade November 1917.

52 LAC RG 9 Vol 4833 1st Division Report on Passchendaele Ridge Operations November 4-12 1917.

53 Radley, p.370.

54 McCulloch, pp.168-169.

55 LBD September 11th 1916.

56 LBD August 7th 1917.

57 LBD December 12th 1916.

58 LBD November 16th 1916.

59 Ibid.

60 LBD September 1917.

61 LBD August 7th 1917.

62 LBD August 7th 1917.

63 LBD August 8th 1916.

64 LBD August 8th 1916.

65 LBD July 17th 1917.

66 LBD March 27th 1917.

67 Ibid.

68 LBD November 4th 1917.

69 LAC RG9IIID3 Vol 4872 WD 2nd Canadian Brigade November 1917.

70 LBD August 2nd 1916.

71 LBD August 11th 1916.

72 Radley, "Ducimus" p.143.

73 LAC RG9IIID3 WD 2nd Canadian Brigade Lessons Learned from the Battle of Hill 70 September 5th 1917.

74 Ibid.

75 LAC RG9IIID3 Vol 4872 WD 2nd Canadian Brigade November 1917.

76 Ibid.

77 Ibid.

78 Ibid.

PHOTO COURTESY OF AUTHOR, WITH KIND PERMISSION OF MR. GLENLYON CAMPBELL

Lieutenant-Colonel Glen Campbell,
Officer Commanding the 107th Battalion, CEF.

LIBRARY AND ARCHIVES CANADA PA-1370

The Modern Major-General. Sir Arthur William Currie.

Generals (unknown), Arthur Currie, Victor Odlum, and David Watson.

Planning the Next Move. Lieutenant-General Arthur Currie discusses his plan of attack at Lens and Hill 70, August 1917.

LIBRARY AND ARCHIVES CANADA PA-007580

Brigadier-General George Tuxford and
Major-General Frederick Loomis share a word in London.

COURTESY OF AUTHOR. FROM THE BLACK WATCH REGIMENT ARCHIVES P012-1

A Contemporary autographed portrait of Major-General Frederick Loomis.

COURTESY OF AUTHOR. FROM THE BLACK WATCH REGIMENT ARCHIVES P012-2

Major-General Loomis stands at the center of his happy headquarters staff.

LIBRARY AND ARCHIVES CANADA PA-002619

Major-General Archibald Macdonnell seated at the centre of his divisional staff.

The cap badge of the 107th "Timberwolf" Battalion, CEF.

LIBRARY AND ARCHIVES CANADA PA-1574

One of the few shots of Lieutenant-General Arthur Currie mounted on horseback.

LIBRARY AND ARCHIVES CANADA PA-4254

Major-General David Watson, centre, in distinguished company. Left of him is Lieutenant-General Arthur Currie, while standing just behind is Major-General Harry Burstall, GOC 2nd Canadian Division.

LIBRARY AND ARCHIVES CANADA PA-6315

Lieutenant-General Richard Turner VC.

LIBRARY AND ARCHIVES CANADA PA-2116

The Journalist-Soldier. Major-General David Watson.

LIBRARY AND ARCHIVES CANADA PA-034110

Major-General A.G.L. McNaughton.

LIBRARY AND ARCHIVES CANADA PA-004543

A typical Canadian Divisional HQ dugout during the Battle of Amiens, 1918. Quite a contradiction to the myth of the General's chateau lifestyle.

COURTESY OF AUTHOR

The Lille Gate at Ypres. The city was the lynchpin of the entire northern sector of the Western Front and subject to many years of direct fighting.

LIBRARY AND ARCHIVES CANADA PA-001137

Canadian soldiers, German prisoners, and wounded move past battle wreckage on the Menin Road.

LIBRARY AND ARCHIVES CANADA PA-001189

Canadian soldiers observing towards Hooge, 1916.

Canadian soldiers searching German prisoners captured after Vimy Ridge.

COURTESY OF AUTHOR

Sanctuary Wood following the great German bombardment.

CHAPTER 5

Major-General David Watson

A Critical Appraisal of Canadian Generalship in the Great War

PATRICK BRENNAN

David Watson's career as a field commander spans the entire war. The combination of senior rank in the pre-war militia, business and social connections in the Anglo-Quebec establishment, and stalwart service to the Conservative party ensured he'd receive a battalion command at the outbreak of the conflict. But what was not guaranteed was his superb leadership of the 2nd Battalion at Second Ypres. A brigade command followed, and after only eight months, selection to command the newly forming 4th Infantry Division, which he subsequently led in action from the later stages of the Somme through the Armistice. Throughout the war, the dapper, ambitious and well-to-do Watson played the intricate politics of career advancement superbly. His selection to command the 4th Division owed as much to his well-cultivated friendships with Max Aitken and Sam Hughes as any military potential. But when the time came, Watson broke with Hughes to emerge as a strong supporter of Arthur Currie. As a military commander, Watson's personal courage was never in question but in some corners his judgement was. Fairly or unfairly, Watson and his division remain forever associated with some of the most costly setbacks suffered by the Canadian Corps – the infamous raid six weeks before Vimy Ridge, the subsequent slaughter on Hill 145 in that attack, the ill-starred assault on Lens, and the bloody setback suffered at the Drocourt-Quéant Line. In battle, there is a narrow line between admirable aggressiveness on one hand, and glory seeking and foolhardiness on the other. Watson's legacy is to be deemed neither a great general nor a poor one, but rather one who was good enough.

When the call to arms went out in August 1914, David Watson was a 44-year-old happily married father of three daughters. At 5'11" he was tall for the times, with a lean, athletic build, dashing moustache, piercing blue eyes and an engaging smile. Watson's profession was journalism, but the business side of it, where journalism paid – in his case as managing director of the *Chronicle*, Quebec City's English-language daily. Watson had done well financially, and he and his family enjoyed a very comfortable lifestyle. "Journalism" was a broad term in pre-Great War Canada, and politics, at least of the behind-the-scenes variety, would just as accurately have described Watson's profession.[1] Both Watson and his newspaper were pillars of the Quebec wing of the Conservative party – no small advantage to an aspiring military officer in the Canadian Expeditionary Force. While many other business and professional men flocking to the colours had impressive political connections, none destined for senior command on the Western Front could match David Watson's, or his skill in utilizing them. Like most members of the Anglo-Canadian male elite of his generation, Watson was immersed in the prevailing ideology of "manliness and militarism,"[2] was an imperialist and was a militia officer. A man in David Watson's position could have cheered others on – a man of David Watson's temperament and mindset had to enlist.

The infighting for senior posts at Valcartier camp was almost as fierce as the war for which the would-be commanders were training, and Watson was in the thick of it. The political implications were such that the Prime Minister even had to intervene on occasion, apparently in one instance on Watson's behalf.[3] With a competent militia record to establish his *bona fides*, not to mention a long-standing friendship with Sam Hughes, Watson's cause was bound to advance. When the first contingent sailed for England in late October, Lieutenant-Colonel Watson was the commanding officer of the 2nd Infantry Battalion.[4]

Watson and his men first saw action in April 1915 at the Second Battle of Ypres. By all accounts, he acquitted himself very well during that bloody three-day seesaw engagement when not a few other senior officers displayed a lack of coolness and competence under fire.[5] Watson was proud of his performance, but devastated by the price his men had paid – two-thirds of his battalion were casualties.[6] He had even carried one of his own wounded out on his back.[7]

Promising battalion commanders were bound to be noticed, and promising battalion commanders with gilt-edged political connections even more so. To enhance his chances, he became an intimate of the cabal of Hughes cronies – principally Max Aitken and Major-General J.W. Carson – who, with Sir Sam's endorsement, effectively ran Canadian army operations in England during the first two years of the war.[8] In August 1915, when Hughes wisely decided to replace Brigadier-General J.P. Landry, a weak choice but the senior French Canadian officer available, as commander of the new 2nd Division's 5th Brigade just before they embarked for France, Watson got the nod – partly on merit and partly by the subtle use of connections. The 2nd Division's commanding officer, Major-General Richard Turner, was a fellow Quebec City resident and friend, which augured well.

The 2nd Division entered battle in the infamous St. Eloi craters fight in April 1916, and was very roughly handled by the Germans, though Watson's 5th Brigade was not directly involved in the worst of it.[9] Watson added to his growing reputation as a "fighter", however, displaying his usual enthusiasm for the attack, much to the disgust of at least one of his battalion commanders, Lieutenant-Colonel J.A. Gunn. The latter, who had absorbed a dressing down from Watson for showing insufficient aggression and questioning his brigadier's orders, threatened to take the matter before their superiors. Clearly hoping he was in line to succeed Turner, if, as was widely assumed, the latter was cashiered for his muddled performance at St. Eloi, or alternatively to get the new 4th Division, Watson recoiled at the prospect of any controversy, and resolved matters with Gunn behind the scenes.[10] In the end, Turner's career as a field commander was saved – for a time – and so was Watson's reputation.

Hughes had decided in early 1916 to send a fourth division to France, and there had been much talk that the Minister's son Garnet would get to command it.[11] Garnet's only rivals were Watson, whom General Alderson considered the best available brigadier among the Canadians, and Henry Burstall, the Corps' chief gunner.[12] Hughes was under growing pressure to clear up the chaos in training and other army operations in England for which he, of course, was largely responsible, and at the same time was desirous of securing the 4th Division for his son. To resolve both, he hit upon the solution of establishing a new "council" in England responsible for running the army's affairs, with Watson as its head as a sort of "inspector general" of Canadian forces in Britain. Watson had a proven record in the field; was respected by at least most of his officer colleagues, including Currie, as well as by the British and Prime Minister Borden; and his loyalty to Hughes, judging by his endorsement of the Ross rifle, the Minister's litmus test for trustworthiness, was a given.[13] Watson was sufficiently professional to have strong views on the need to reform army affairs in England, and agreed to serve in some capacity, but only until the 4th Division was his. Hughes's agent in the negotiations – Sir Max Aitken – either didn't appreciate this, thought Watson could be outmanoeuvred, or was misled.[14] In fact, Watson had the stronger hand, for regardless of Canadian opinion, the British would never accept Garnet Hughes as a divisional commander. By early April, Sam Hughes and Aitken had promised Watson the 4th – the actual power of appointment lay with the War Office – and Watson had reluctantly agreed to take charge temporarily of all Canadian troops in England, subject to having "sole control of all military matters,"[15] namely the removal of senior incompetents from their posts. Whether he concluded that the chances of anything that Hughes organized actually working out were hopeless, or realized that the Minister's influence was waning, Watson, his coveted 4th Division in hand, quickly abandoned the council.[16]

Watson enthusiastically set about assembling the units for his new command. The War Office seconded a very able, and equally ambitious, young staff officer to serve as his GSO1, Lieutenant-Colonel Edmund Ironside. For the next twenty months, the two would shape the 4th Division in a working relationship, which would be mostly productive but at times problematic for both the division and Watson's reputation. The 4th Division represented the last opportunity for infantry battalions raised in Canada to reach the Front as intact units, and avoid the fate of being broken up for reinforcements. All battalions had been raised with the implicit prospect of service in France front and centre, not just for their commanders, but for their officers and men as well. Watson and Ironside had their pick of the litter, and the competition – and "political" infighting – to be selected was fierce. Watson felt he had extracted a pledge from both senior Canadian and British officials that he would have a free hand in organizing the division – he was to be disappointed.[17]

Watson and Ironside scoured the training camps of England for suitable battalions. The prospective composition of the division shifted almost daily, not the least because of Hughes' continued meddling. The Minister's "suggestions" were driven by his usual priorities – domestic political considerations and his attitude toward a battalion's commanding officer. Watson, however, took his job very seriously. While hardly averse to factoring in "political" realities, he'd learned in the field that a commanding officer needed complete authority to appoint his own subordinates and fought hard to get his way.[18] When Hughes, Carson and Aitken pushed for an all-Western 10th Brigade, conveniently mimicking the make-up of Watson's former 2nd Brigade command, Watson and Ironside agreed, but the allocation of suitable battalions for the other two brigades proved more complicated.[19]

Aitken initially suggested the 53rd (Saskatchewan), 54th (British Columbia), 74th (Ontario) and 75th (Ontario) Battalions for the 11th Brigade, and the 51st (Alberta), 72nd (British Columbia), 81st (Ontario) and 87th (Quebec) Battalions for the 12th.[20] The 45th (Manitoba), 65th (Saskatchewan), 66th (Alberta), and 77th and 84th (both Ontario) were subsequently suggested, and dutifully inspected. Looking for battle worthiness above all else, neither Watson nor Ironside always liked what he saw on the parade square, and most of these units failed to pass muster.

When Watson learned that Lieutenant-Colonel Peers Davidson had adamantly refused to have his 73rd broken up – like all Montreal battalions, it had influential patrons in politics and business among the city's powerful Anglophone elite, and there were almost irresistible pressures to ensure such battalions made it to the Front – he was ecstatic. It "is a fine regiment, well officered and well disciplined and would be a great acquisition ...," he enthused.[21] As for the 72nd and 87th Battalions, "both are splendid," he wrote after reviewing them at Shorncliffe.[22] But the 87th posed another problem. Neither Watson nor Ironside had confidence in its inexperienced 31-year-old commander, Lieutenant-Colonel Irving Rexford, who had only recently been promoted from second-in-command, and happened to be the son-in-law of General Carson. Carson wanted the battalion and its incumbent commander both taken, but with Hughes not caring, Watson reminded the General that he intended to "report very straight on the matter."[23] In the event, he got to keep the 87th and with a new lieutenant-colonel to run it, though the man was Carson's, not his own, choice.[24] Watson settled on the 102nd (British Columbia) after an impressive performance during on one of his many inspections – "by far the best" of four he'd seen that day – and similarly the 78th (Winnipeg Grenadiers).[25]

By early summer, the 4th Division's composition was settled, but the delays in getting it done – certainly none of it Watson's fault – appreciably slowed training.[26] Watson had done the best to ensure merit prevailed in the selection of his infantry units, though at some considerable risk of alienating various interests inside and outside the army.[27] His considerable diplomatic skills had been most useful in fending off the intervention of Hughes and his associates,[28] but the process of battalion selection was overseen by the War Office as well, no doubt wisely given the disorganization and cronyism which passed for Canadian military "administration." The assistance of Max Aitken, who was a confidant of both Hughes and the British, had been enormously helpful in both regards. The Canadian general had gone out of his way to cultivate Aitken's support and during the formation of the 4th Division it paid handsome dividends for which Watson was most grateful.[29] That said, working intimately with Aitken had also convinced him that the press baron and fixer-extraordinaire was completely untrustworthy,[30] though these feelings he kept to himself. After all, Aitken might be an unreliable friend, but better an unreliable friend than an enemy, and in any case he might prove useful in the future.[31]

At a time when the Canadian Corps was crying out for infantry reinforcements – the army suffered 20,000 casualties in June 1916 alone – Watson's attempts to keep the best-trained reserve battalions in England for his 4th Division struck some of his contemporaries (and not a few historians since) as a misguided sense of priorities or worse still, self-serving ambition.[32] In fairness, the decision to give the 4th Division's formation priority, then repeatedly delay decisions on its make-up, which resulted in tying up so many battalions in an administrative "no-man's land", was not Watson's though he certainly concurred with it. Nor was he responsible for the dearth, not of men but of *trained* men in England, which slowed reinforcements to a trickle in the late spring of 1916. Hughes and Generals Carson, Steele and MacDougall can claim that honour. But by mid-month, thanks to the War Office's direct intervention, Watson had to stand by and watch months of hard work undone as successive drafts of his picked men were shipped to France.[33]

If Watson had been largely successful in assembling the battalions (and battalion commanders) he wanted, the selection of his brigadiers would be another story. Fancying himself as a picker of generals, and much more likely to know and hence like (or dislike) the personalities involved, Hughes was less willing to give ground on these choices. When Watson's initial suggestions reached Carson, the latter strongly intimated "that the minister was to appoint same." Watson could only point "out [the] gravity of any change when recommendations are made on merit alone."[34] Undeterred, Hughes settled on three men who represented the entire spectrum of talents available: Frederick Loomis, a seasoned and very capable Brigadier then commanding the 7th Brigade; Lord Brooke, a former British regular officer and Hughes' crony who'd briefly commanded the Canadian 4th Brigade in France in 1915 and subsequently been appointed to run Bramshott training camp; and Hughes' own younger brother, Lieutenant-Colonel St. Pierre Hughes, then commanding the 21st Battalion.[35] Watson accepted Brooke for the 12th Brigade, presumably in an (unsuccessful) effort to avoid Hughes' sibling in the 10th. Loomis would have been entirely acceptable had he not harboured such a low opinion of Ironside whom Watson trusted implicitly. When Loomis departed by mutual consent after a short stint in charge of the 11th Brigade, Watson happily accepted Victor Odlum, recommended by Carson and known to be well regarded by the new Canadian Corps' commander, General Byng.[36] Odlum, a wealthy Vancouver executive, was an experienced battalion commander with a reputation as a "fighter." Meanwhile, Watson countered St. Pierre Hughes with another former British regular, Lieutenant-Colonel Edward Hilliam, who had commanded a battalion under Watson.[37] Try as he might, however, Watson was thwarted. When the 4th Division moved to France in August, Hughes was one of its brigadiers.[38] The fates, however, were on Watson's side. Only a month after reaching the trenches, Lord Brooke was seriously wounded and permanently replaced by James MacBrien, an experienced staff officer who would prove to be an excellent brigadier. Replacing St. Pierre Hughes required that he discredit himself, which he did in due course during the fighting on the Somme, and that Sir Sam's power to protect his brother be eliminated, which the Prime Minister accomplished around the same time by dismissing him from the cabinet.

During the first eighteen months of the war, David Watson had been among the group of senior army officers who broadly supported Sir Sam Hughes.[39] Thereafter

Watson had slowly begun to sever his links. While association with Hughes no longer advanced, but rather hindered, the career of an ambitious officer, supporting the "Hughes way of war" – cronyism and the chaotic, inefficient administration that inevitably followed in train – had become impossible for anyone who was committed to professionalizing the CEF and crushing the Hun. There was no great joy in the diary entry recording the Minister's dismissal from cabinet, just a simple observation that "this will likely be a wise move for all concerned."[40]

By December 1916, Watson was aggressively pressing Byng to fire Hughes, and only reluctantly agreed with the Corps commander's advice to put Hughes on notice with a severe reprimand. A month passed, and Watson pressed harder, accusing Hughes of showing no improvement and severely compromising his brigade's battle worthiness. Byng now agreed, and duly authorized Hughes' dismissal, appointing Hilliam in his place.[41]

The 4th Division had set out for France in August, with the Somme campaign already into its second month. In the way of soldiers which is so incomprehensible to civilians (and military historians), morale was high among officers and men longing for the chance "to hit the Hun for six." Watson was understandably pleased.[42] Through the end of September, the 4th Division was assigned to a quiet portion of line in the Ypres salient. Watson busied himself supervising training – battalion and then brigade-sized practice schemes, and of course the inevitable raiding to "blood" junior officers and men.[43] Apart from a sprinkling of senior officers with battle experience, the division was terribly "green" and their general knew it. In fact there was little time to inculcate more than the basics. He and Ironsides visited the Corps school for officers, noting that "it was very interesting and we got hold of a lot of new ideas." He also paid a visit to his old 5th Brigade headquarters shortly after their attack on Courcelette, "and it made me sick to think of all the officers I knew, who are gone."[44] To familiarize himself with their future battlefield, he also arranged to tour the area around Pozière where the Australians had suffered such slaughter – "Not a foot of ground that was not shelled, and the quantities of rifles, bombs, grenades, equipment, skulls, arms, legs etc was awful…"[45]

Watson's division entrained for the Somme on the night of October 2-3, their destination the Ancre Heights section of the line, where the 1st and 3rd Canadian

Divisions were shedding so much blood to capture Regina Trench. The 4th Division was attached to the BEF's Reserve (soon to be Fifth) Army under the command of Hubert Gough, a former cavalryman and Haig favourite who was already known among the Canadians for his excessive zeal for the ill-planned attack. The Somme had produced mixed results for the Canadians by the time the 4th Division made its first attack – the proverbial thousands of casualties for a few acres of mud. On the eve of battle, Watson considered Odlum's 11th Brigade his most proficient unit, with MacBrien's 12th a bit behind.[46] The 10th Brigade – still under Hughes – remained an unknown quantity. Participating in a joint Anglo-Canada assault on the 21st, Odlum's battalions, with excellent artillery support, took their objective – a section of Regina Trench – in short order and with light casualties.[47] It was an auspicious start for the "Green Patch". The follow-up attack, scheduled for three days later but delayed 24 hours by a steady downpour, was shaped, the CEF's official history suggests, by unwarranted optimism. With tragic irony, the 10th Brigade's plan, which called for a single battalion to seize a section of Regina Trench that had eluded the 3rd Brigade's attack some two week's earlier, was printed on the back of army will forms. When the barrage proved ragged and ineffective, German machine gunners tore gaping holes in the ranks of the Manitobans as they stumbled forward in the mud, killing or wounding nearly half of them. Not a single soldier reached the objective. Hughes later complained he had been told by the artillery commanders that most of their guns would not be registered by zero hour, and that he'd personally informed General Watson, on the assumption the attack would be delayed.[48] Some days after the attack, Watson and MacBrien reconnoitered bits of Regina Trench. "It was desperate hard work and often we were up to our middles in slush and slime," he recounted in his diary. "I was absolutely covered with mud and soaked to the skin [and] saw large numbers of unburied bodies all around in all sorts of positions."[49]

The weather now deteriorated, with almost continuous rain, and further attacks were planned and repeatedly called off. Still, to impress the French at an upcoming conference and buck-up the Romanians, Haig was determined to make one more push – and Gough was willing to comply.[50] As a prelude, the 4th Division was ordered to prepare an assault on the remaining German-held portions of Regina Trench. The night attack by Odlum's men succeeded brilliantly, again thanks largely to a devastating artillery barrage, which earned kudos from Watson.[51] The ensuing British attack was also a success, which unfortunately inspired Gough to mount a follow-up in which the 4th Division's 10th and 11th Brigades (plus a single battalion from the 12th) were to take Desire Trench a mere half a mile north of Regina Trench. As usual, Gough's preparations were rushed and poorly coordinated, and the weather was appalling, even by Somme standards. The attack was launched on November 18, and with visibility impaired by snow, sleet and rain, the infantry got lost and the gunners couldn't see their targets. Although things initially went well enough for Odlum's Brigade, they immediately fell apart for Hughes' force and the British brigade on Odlum's flank, and by nightfall, after the Canadians had suffered 1250 casualties, the Corps commander called off the attack.[52] It had been a difficult initiation. Several 4th Division attacks had partially or completely failed, yet under the most appalling conditions, others had gone very well, with the 11th Brigade showing particular promise. A divisional commander was not an autonomous actor, and most of the problems his men encountered were certainly not attributable to Watson's miscalculations. For an inexperienced formation, and a divisional commander – and chief staff officer – in their first battle, the results were acceptable.

The bleak winter months following the Somme were a time of introspection for the British Expeditionary Force and its Dominion formations. Out of the blundering and carnage, many lessons had been absorbed by more insightful minds in the various commands and now, under the direction of GHQ, these were formulated into new tactical doctrine – the genesis of the "bite and hold" attack.[53] Nowhere was innovation more actively pursued than in Byng's Canadian Corps. A fortnight before the 4th Division's attack on Desire Trench, the Corps' BGGS, Brigadier-General Percy Radcliffe, sent out a memorandum to all four divisions canvassing:

> ...lessons to be derived from [recent] operations ... in order that the valuable experience gained ... by the Corps may be turned to the best account in future operations. Any points of interest in connection with tactics, organization and administration would be dealt with and proposals submitted for improvements on the methods adopted.[54]

Watson, Ironside and their subordinates made their contributions, though on one of the major questions – the re-organization of the fighting platoon into a self-contained formation of specialist sections ultimately adopted throughout the BEF – both men felt nothing was to be gained by change.[55] Lest Watson be dismissed as blindly orthodox, Louis Lipsett, who was generally viewed as second only to Currie as a divisional commander in the Corps, wholeheartedly agreed with Watson's assessment.

As it had in the case of Brigadier Hughes, combat, and the preparation for combat, revealed much about the strengths and weaknesses of any level of military leadership. By year's end, Watson had arranged the replacement of three of his twelve battalion commanders, two of them from the 10th Brigade. All three had performed inadequately and suffered severely from the strain of fighting their battalions. Watson proved to be ruthless in this regard, having absorbed the lesson as a battalion commander that displays of "humanity" toward poorly performing officers came back to haunt their commander and kill their men.[56]

By late winter, Watson, like the rest of the senior officers in the Corps, was putting in long hours preparing for the forthcoming attack on Vimy Ridge, preparation which culminated in the great victory of Easter week. Raiding had become a Canadian mania by early 1917, and all four divisions, actively encouraged by Corps headquarters, staged these small-unit attacks aimed at seizing control of No Man's Land, terrorizing the enemy, gaining valuable intelligence (at least about German front-line defences) and cultivating "the attacking spirit."[57] Watson and Ironside, having gained success with smaller raids, decided to launch one on a grander scale, primarily relying on gas rather than an artillery bombardment to neutralize the German positions being attacked. Their objective was Hill 145 on Vimy Ridge. It was clear from the plan that neither officer understood the practicalities of offensive gas warfare. How poison gas, being heavier than air, was to creep uphill and overcome the German defenders escaped them. Furthermore, preparations to launch the gas attack were such a major engineering undertaking that they could not be hidden from the enemy, thus losing surprise. Careful training – the Corps' mantra – was also ignored. When the commanding officers of two of the four battalions committed to the attack – Lieutenant-Colonels Arnold Kemball and

Sam Beckett – aired their doubts to Odlum, he, too, became alarmed and carried their anxieties to Ironside, who dismissively brushed them aside.

The raid went ahead on March 1 – disastrously. When the gas failed, the attackers were slaughtered. Casualties – including both Beckett and Kemball killed – were 687, or just under one-half of the assaulting force. For no gain, one-third of the division's infantry battalions had been gutted barely a month before a major attack. MacBrien and Odlum were aghast, the latter particularly furious that the raid had gone ahead when it was obvious from heavy rifle fire that the gas had had little or no effect on the German defenders.[58] Watson was stunned by the day's events, simply noting with staggering understatement that "to our surprise the gas did not have the effect anticipated and the Boche[s] were waiting for us."[59] The day following the attack, he had thoughtfully sent Odlum to London to personally convey the news of her husband's death to Mrs. Beckett.[60] While Watson seems to have felt genuine grief over the heavy losses among his officers and men, he doesn't seem to have felt any personal responsibility, at least beyond that shouldered by any commander who had done his best in preparing an attack which failed.

Certainly there was an unreal quality about the planning for the raid. As Cook concludes:

> Watson and Ironside overlooked the very real and documented failures of gas because they wished it to succeed … and [hence] succumbed to that strange act of delusion that occurs when the odds are stacked against you and caution gets thrown to the wind.[61]

In analyzing Watson's role in the debacle, he has suggested that the underlying problem was that the "supremely self-confident, forceful and opinionated" Ironside overawed the general to the point of intimidation. There is at least substantial indirect evidence that this occurred, at least early on in Watson's command. As Aitken had noted, when pressured by superiors, or strong men, he could be indecisive, not the best attribute in a field commander.[62] Furthermore, Watson, though dashing and self-confident in public, was a remarkably insecure man who appears to have longed for his superiors' approval. In a diary entry only weeks before the attack he recorded poignantly "[I] think it strange that Gen[era]l Byng has only been to see me once in 1½ mo[nth]s and then only to unnecessarily severely criticize." Sixteen days later, when Byng visited his headquarters, Watson gleefully recorded that "he was greatly pleased with the Division and all the work and raids that had been carried out (author's emphasis)."[63] Watson oversaw the broad planning for the raid, met with his staff officers and brigadiers to discuss it, and at one point even threatened to call it off if sufficient artillery support wasn't guaranteed, evidence that he considered he was exercising due oversight.[64] Unfortunately, his assessment of what would be "sufficient", like so much else, fell short of the need, one of the after-action reports pointedly emphasizing that the barrage was "not sufficiently concentrated and caused no slackening of enemy fire."[65] But what Watson thought he was doing, and what he was doing, were two different things. The fact is preparation for the raid was haphazard and revealed inflexibilities in the division's operational culture, particularly a failure to heed the warnings of experienced front-line officers.[66]

Thirty-nine days after the ill-fated March 1 raid, Watson's men stormed Vimy Ridge again. For that attack, the 4th Division was allotted the most difficult task, capturing the heavily defended heights of Hill 145 and "The Pimple", and was expected to encounter the hardest going. In the event, the assault battalions from Odlum's 11th Brigade were all but annihilated in the opening stages of the attack, suffering hundreds of killed and wounded, and those of MacBrien's 12th also suffered very heavily. Indeed, for a time, the outcome of the 4th Division's attack seemed to hang in the balance. Hill 145 was finally stormed on April 11 after a succession of very costly attacks had worn the Germans down, while "The Pimple" was gained a day later. Watson and his staff appear to have learned the obvious lessons from the March raid. The infantry were thoroughly trained in the new platoon-fighting tactics, to which Watson became a great convert.[67] More importantly, they had drawn up plans for their supporting artillery to crush German defences under a hurricane of metal and high explosive. But a lone area in the path of the 11th Brigade's attack was not hit – inconclusive evidence suggests that Odlum, presumably without seeking approval from divisional headquarters, may have cancelled the barrage on that section of German trench in the hopes of capturing it intact for use as his forward headquarters.[68] On the morning of April 9, machine gun and rifle fire pouring from those undamaged German positions had stopped the Canadians dead in their tracks. Despite this appalling miscalculation, Odlum was a competent brigadier, admired by Watson, Byng and Currie, and his career suffered no damage. Instead, responsibility for the error was placed on the shoulders of Major Harry Shaw, the commanding officer of the 87th Battalion, whose career and reputation were both ruined.[69] As was inevitably the case during the Great War, the growing chaos and collapse of communications rendered Watson's task of controlling the opening stages of the attack hopeless. Although the battle remained a bloody *mêlée*, from the second day onward his role became clearer, and his performance revealed much coolness and sound judgement.

In the aftermath of Vimy, Watson had three more battalion commanders replaced, including the aforementioned Shaw. The other two were Lieutenant-Colonel Charles Worsnop, an officer with a fine record who'd temporarily taken over the 75th when Beckett was killed, but was worn out and no longer measuring up according to Odlum,[70] and Lieutenant-Colonel William Winsby of the 47th. Acting on numerous negative reports from Brigadier Hilliam, Winsby had been under Watson's orders to "make good" since the middle of March. Watson's patience had now run out. And with the disbanding of the 73rd Battalion as part of a reinforcement reorganization, he had also lost Lieutenant-Colonel Herbert Sparling.[71] Knowing what it was like to have men around him who weren't his own choices, Watson trusted the judgement of his three brigadiers sufficiently to accept their advice on replacements, although the ultimate decision was Byng's.[72] After six months at the Front, Watson was most satisfied with field commanders. Hilliam, Odlum and MacBrien were all capable, battle-experienced men whom he got along with well, while of the battalion commanders, half had been selected on Watson's watch and the six he'd inherited with their battalions in 1916 who still held their posts had all proven to be sound men.

The major development following Vimy was the departure of Lieutenant-General Julian Byng for Third Army. Byng had inherited a force whose senior officer corps was faction-ridden, politicized and included many who were promoted beyond their level; whose administration was plagued by inefficiency; and whose performance on the battlefield was mediocre and showing few signs of improving. Barely a year later the Canadian Corps had been transformed into a first-class fighting formation characterized by its professionalism and *élan*.[73] The mentoring of senior officers had been one of Byng's (and Radcliffe's, though the BGGS would stay on for another six months) singular contributions. Watson had come to admire Byng, not the least for this reason, and he made sure that the 4th Division staged an appropriate send-off when the outgoing Corps commander paid them a last visit on June 8. "He was much affected," a pleased Watson recorded in his diary.[74] The departure of Byng had been mooted for several months. A friend of Lieutenant-General Turner – they both hailed from Quebec City – Watson had supported the latter's transfer to England in November 1916, offering that "he will make a splendid appointment as he understands fully our requirements" – namely major reforms to administration and training.[75] Many others saw Turner's "promotion" as more a way of removing a mediocre performer from field command, which it was, and to keep him from being in line to succeed Byng.[76] The commander of the 1st Division, Arthur Currie, was the obvious alternative to Turner, and much favoured by the British.[77] In early February, two months before Vimy Ridge, and with Byng's approval, Currie began to solicit opinions from the Canadian divisional commanders as to a strategy they might collectively follow when the time came to ensure their views would be heard. When the commanding officer of the 2nd Division, Henry Burstall, approached Watson to attend a meeting at Currie's headquarters, Watson balked – fearing perhaps the "political" optics. Instead, he invited Currie to 4th Division headquarters where the two men "had a long chat and [I] gave him my views on the Canadian officer question," presumably along the lines of what he'd already said to Burstall, namely that his guest was "one of our big, capable and efficient officers [who] would compare favourably anywhere."[78] Knowing full well he wasn't in the running, Watson's position was straightforward. The job must go to a Canadian – and since Currie was the best Canadian, it must go to him.[79]

When Currie finally got the plum in June, the fear that his pre-war embezzlement scandal would become public threatened everything. It was the willingness of the wealthy Odlum and Watson to loan him the money to clear his debt that saved Canada its greatest military commander and a damaging leadership crisis in the Corps.[80] Watson's chief motivation was to save Currie's career – it was the action of a patriot committed to ensuring Canada played its full part in the war, and there is no evidence to discredit this assessment. He had also now gone full circle – from an ally of Sam Hughes to a fierce opponent, from a brilliant young commander in

Hughes' eyes to someone who could be peremptorily dismissed as "unfit".[81] Currie's promotion to Corps commander and Hughes' exit from the cabinet didn't mean the end of politics in the Canadian Corps – it merely changed its nature. This was not the old regime re-cast – merit counted now. Still, until the end of the war, Currie's position was never really secure – the pre-war scandal continued to hang over his head, even if the debt had been repaid, and could still have brought him down. The embittered Hughes' faction was a constant threat against which Currie needed allies, and he remained deeply suspicious of anyone he felt he couldn't trust on the "Hughes" issue.[82] If he didn't prove himself incompetent, Watson's position was now secure.

The Battle of Hill 70/Lens, the Corps' next major engagement, and the first directed by Currie, did not go well for the 4th Division. The Canadian attack, which after several delays because of poor weather, was intended to draw German reserves away from the Passchendaele offensive. The story of the Canadian diversion is well known – told to attack the coal-mining town of Lens, Currie, after personally reconnoitering the terrain, instead insisted on capturing the dominating heights of Hill 70. In a textbook "bite and hold" operation, the 1st and 2nd Divisions successfully stormed their objective on August 15, then repulsed no fewer than twenty-two German counter-attacks intended to wrest back the hill, inflicting very heavy losses on the enemy.[83] It was a stunning victory, the most one-sided won by the Corps till that point. Unfortunately, it didn't end there. Urged by the British to press on into Lens, Currie seems to have suspended his better judgement and gone ahead with an attack, despite sufficient intelligence that the ruined city was heavily defended, an urban wasteland riven with strong points and tunnel systems which guaranteed any assault would quickly degenerate into a close-quarter "street fight" where Canadian advantages in artillery firepower would be largely negated. It fell principally to Watson's 4th Division – chiefly the four battalions of his Hilliam's brigade – to carry out this subsidiary plan, which it attempted over four days of hard and very costly fighting. Currie finally called a halt on August 25, but not before the 4th Division had incurred 4,000 casualties including an entire battalion wiped out trying to capture and then hold a coal slagheap called the "Green Crassier."[84] The lack of time for specific training, the general absence of proper intelligence (and the ignoring of what sound intelligence was gained), a sense of overconfidence (which in the warm afterglow of Hill 70 and Vimy Ridge must have pervaded the entire command), and a general sense of "hurry" all contributed to this doomed enterprise. But these can mostly be attributed to poor judgement and mistakes at Corps Headquarters rather than being laid at Watson's and Ironside's door. After interviewing his officers and surveying the battlefield, Watson, with Ironside's input, prepared a report outlining what had gone wrong, and at least indirectly took responsibility for some of the miscalculations and miscommunications, as he should have. Hilliam, too, bore a share of the blame, particularly for the Green Crassier fiasco, since Watson's practice was to give his brigadiers considerable leeway in running "their" battles – in this instance, perhaps too much leeway.[85]

With the rest of the Corps, the 4th Division moved on to the Ypres salient and Passchendaele where physical conditions were the worst Watson's men had yet encountered. Nevertheless, the 4th Division executed their part of Currie's attack about as flawlessly as could have been achieved under the circumstances, and fulsome congratulations were offered by both Currie and General Plumer.[86] Watson had always been a physically active commander – not just visiting his brigade and battalion headquarters but surveying the front lines. Now he spent much more time attending to detail such as pouring over barrage plans and the like, and it seemed to be paying dividends. He had always displayed an ability to learn, and clearly he was maturing as a commander.

By Passchendaele, Watson had soured on Brigadier-General Hilliam – the latter had strengths, but handling men wasn't one of them. As a result, his relations with divisional staff and his own battalion officers were constantly strained. Reading between the lines, it appears that Hilliam might have been "too English" in his approach, always a potential source of difficulty when British officers worked with Canadians. After Watson gave him the option of mending his ways or transferring to the British army, a stern lecture from Currie decided the issue. He was replaced by Brigadier-General Ross Hayter, who, while raised in Canada was a British regular, and had been serving with the Canadians in a staff capacity, most recently as GSO 1 of the 1st Division.[87] Currie knew that Hayter was about to be recalled for a brigade command in the BEF, and as "an extremely valuable officer ..., the Corps could very ill afford to lose him."[88] Watson was agreeable. Hayter commanded the 10th Brigade almost through the end of the war, and proved superior to Hilliam in all respects.

December 1917 was a bleak time – "Thoughts of home and all those dear ones there keep sweeping over me all day long," Watson confided to his diary. "It is a bitter cruel thing this war."[89] He passed the time inspecting units and attending conferences. A gregarious man, most evenings he socialized with his officers when he wasn't buried in paperwork. The camaraderie seemed to keep his spirits up. So did the Unionist election victory, Watson proudly recording that his division had provided 17000 votes and probably "not 5 percent for the opposition."[90] By year's end, Ironside, too, was gone, replaced by Brigadier-General Édouard Panet. A gunner from the pre-war Permanent Force, Panet, at 36, was eleven years Watson's junior. In 1913, he'd completed the British staff college course at Camberley. Since professionally trained staff officers were almost non-existent in the Canadian Expeditionary Force, Panet had served exclusively in that capacity. He had been with the 4th Division from the outset, first as Assistant Adjutant and Quartermaster General, and latterly as a "staff learner" under Ironside.[91] Watson had relied heavily upon Ironside for both practical military advice and counsel – not a few would have said relied too much. But at any rate in Watson's eyes they were a successful team. At least in 1918 no one would be able to say "Tiny" Ironside ran the 4th Canadian Division.[92]

The final year of the war began with one of the acrimonious "flare-ups" with a subordinate officer that seemed to occur periodically in Watson's command. In this instance the disgruntled officer was Lieutenant-Colonel John Warden who'd raised the 102nd (North British Columbia) Battalion in 1915 and had commanded them ever since. Warden was an excellent battalion commander, tough but fair, aggressive, a plain-talker, and by all regards, much loved by his men. But he was the sort who rubbed others the wrong way, mistook ambition for falseness, and generally detested the "political" skills that Odlum practiced so effortlessly. Warden and Odlum hardly spoke, and the former finally requested a transfer out of the Canadian Corps. But Odlum refused to forward the application, according to Warden because Currie would have had to order an investigation, the results of which would have caused much embarrassment all round. Instead, Warden went to Currie on his own, accusing Odlum and Watson of being "mercenary men" bent on using their commands to gain fame. Odlum was the cleverer of the two, "a most avaricious decoration hunter," bent on gaining Watson's job with the latter "not smart enough to know it."[93] Currie seemed unmoved by the incident, and how widely shared Warden's views were within the officer ranks of the 4th Division remains unknown.

The Canadian Corps spent the late winter and spring of what turned out to be the last year of the war holding defensive positions and carrying out training for semi-open warfare and the offensives they hoped would come. The 4th Division was not engaged in stemming the German spring offensives, though the possibility of action during the grim weeks of late March and early April was never far from anyone's mind. During one visit to Odlum's headquarters with Panet, Watson joked darkly that "we all had some revolver practice" just in case things got really bad.[94] Waiting, however, did not mean wasted time. Watson kept his officers and men on their toes, carrying out a particularly busy schedule of visits to his brigades and battalions to monitor their preparations and generally maintain a high level of preparedness and spirit. He clearly embraced Currie's notion that the greatest impact generals could have on a Great War battle occurred during the months and weeks before their troops engaged the enemy, when under their direction subordinate commanders and staff officers could exhaustively prepare the men and plan for every contingency.[95] A perusal of Watson's schedule from mid-May through mid-July shows him immersed in such activities, with 31 entries listing attendance at training exercises and tactical schemes staged by his infantry battalions or artillery or engineering brigades or to have discussions with or hold briefings for his brigadiers and battalion commanders.[96]

The definitive verdict on Watson's generalship would be written during the great battles of the Last Hundred Days. He learned from Currie only on July 29 that the Corps was to be part of "a real big show."[97] On the 1st, he briefed his brigadier-generals on the broad outlines of the plan, then moved with his headquarters secretly to the Amiens area on August 3, the mass of the soldiery and their equipment following shortly thereafter. The next day Currie personally telephoned Watson with an outline of the 4th Division's role, then assembled his Divisional commanders for a detailed briefing on the attack plan after which Watson in turn briefed his divisional artillery commander, Brigadier-General William King. On August 6, Watson had a long meeting with Currie and his BGGS, Brigadier-General Norman "Ox" Webber, to hammer out further details, then with Panet discussed these matters with Odlum, MacBrien and Hayter as well as Lipsett, with whose 3rd Division he would be co-operating.[98] Amiens was the biggest Canadian operation of the war, where instead of the months to prepare for Vimy, the Corps was given less than a week. Evidence of the smooth functioning of the Canadian army, and the competent leadership of its generals, could not have been more clear.

The 4th Division's role was to move forward from reserve at zero-hour (at first dawn, just after 4:00 am), with Odlum's and MacBrien's brigades side by side, and pass through the 3rd Division shortly after noon, before pressing on to the day's final goal, an advance line captured by British and Canadian mounted troops. Virtually all objectives were in hand by nightfall. Changes in Fourth Army intentions that night caused great confusion as Canadian plans had to be remade at the last minute, causing the second-day's operations to begin late and providing the enemy with some breathing space to strengthen his shattered defences. Nevertheless, Watson's division again made good progress. On the third day of the operation, the 4th Division attacked again, but this time the 10th and 12th Brigades encountered very stout resistance and suffered heavy casualties. Watson's timely intervention in moving up a substantial force of artillery to deal with German strong points allowed his exhausted battalions to take their objectives, and fend off determined counter-attacks, but the impetus of the Canadian attack was spent.[99] On the afternoon of August 14, while recent events were still fresh, Watson had assembled all available senior officers to go over "the many lessons learned in the fighting so far."[100] Two days later he discussed these at length with the Corps commander.

In a matter of days, the Canadian Corps faced an even more daunting task – breaching the Hindenburg Line between Arras and Cambrai, specifically the formidable Drocourt-Quéant position. It fell to the 2nd and 3rd Divisions, in a series of preliminary attacks commencing August 26, to wear down the German defences and gain proper jumping-off positions for the main assault scheduled for September 2 which would hurl the 1st and 4th Divisions at the Drocourt-Quéant Line proper. Currie's plan, as broadly envisaged by Haig's headquarters, called for the Canadians not only to break through the Drocourt-Quéant defences, but rapidly press on toward Cambrai, and in particular seize crossings over the

Canal du Nord.[101] Having embraced risky attacks in the past, Watson was nonetheless taken aback when he and Panet were briefed on August 29, noting in his diary that "it is a very ambitious programme and I doubt if it can be carried through to the extent they have laid down..."[102] Daniel Dancocks, Currie's hagiographic biographer, criticizes Watson (as well as the counter-battery staff officer, Lieutenant-Colonel Andrew McNaughton, who also harboured doubts about the plan) for not raising their concerns.[103] Given Currie's reputation for encouraging candid exchanges at such briefings, it is quite possible that they did raise objections, but that these were challenged and dismissed. The key to breaking through the Drocourt-Quéant Line lay in storming the Mont Dury sector that dominated the ridgeline on top of and behind which the German army was dug in.[104]

Panet and an ailing Watson, who, like many of his officers and men, was suffering from dysentery, briefed the 4th Division's brigadiers late on the afternoon of August 30. Divisional intelligence concluded that the enemy would "hold their ground and fight a determined defensive battle if necessary."[105] The original plan drawn up at Watson's headquarters called for MacBrien's brigade to attack over the ridge and seize its Red Line objective on the far slope. Both remaining brigades would then pass through to create and exploit a breakthrough. Despite continued injunctions from Currie that the 4th Division "must get through," last-minute changes that were forced on Watson halved the follow-up force to merely Odlum's brigade.[106]

The final plan called for MacBrien's brigade, now less one battalion, to be carried onto its Red Line objective by an intense artillery barrage. Unfortunately this objective was set at the extreme range for most of the field artillery, save one battery that would be pushed forward as the attack went in. Long-range artillery fire would actually be suspended – even smoke shells were forbidden – in a 1000-yard-wide corridor along the Arras-Cambrai highway to permit Brigadier-General Brutinel's Independent Force[107] to pass through the gap torn by the 4th Division infantry and race ahead to seize a bridgehead across the Canal du Nord (the Green Line, and the 11th Brigade's objective), some 3½ miles distant. Based on past experience, Watson had little faith in deploying the motorized machine gun units independently and en masse in offensive operations.[108] But they promised an easy victory that would save lives later, and Brutinel and his technology had

Currie's ear.[109] All these preparations to facilitate the mobile thrust merely ensured that numbers of German artillery pieces and machine guns would be able to fire unmolested at vehicles, horses, and infantry alike. Against this possibility, orders simply called for such resistance to be overcome "with the utmost vigour."[110] Regardless, to make the dash possible at all, success had to meet the 12th Brigade's attack within three hours. If the Red Line, including Mont Dury itself, were not captured, neither Odlum's battalions nor the Independent Force would be able to proceed, nor would it be possible to bring the supporting artillery forward. The final orders issued to the 11th Brigade made Corps headquarters' optimism clear. The Canadians would "exploit the success, seizing the high ground east of the Canal du Nord." As for any enemy forces between the Drocourt-Quéant Line and the canal, they would "be thrown into confusion by the attack, [cut off] ... and rapidly dealt with" by the pursuing Canadians.[111]

That even the best Corps make mistakes when they are rushed in preparation, faced with overcoming formidable defensive positions, and infused with over-confidence was made clear in the 4th Division's disastrous attack on September 2. MacBrien's force initially made good progress, and soon a substantial part of it had crested the ridge and proceeded down the far side, out of sight. There, his infantry encountered "the most severe machine gun opposition" they'd yet seen. Under withering fire, and with little cover, groups of desperate men managed to advance another 600 yards to the Red Line, which they held to precariously. All this was unknown to Odlum, who'd been given the freedom of action to attack if he had "clear knowledge of what was in front of [him]."[112] Even though he lacked such "clear knowledge," Odlum was chafing to go, but Watson – less certain, more cautious or simply vacillating – delayed the attack for 90 minutes. It made no difference in the outcome. Odlum's infantry finally advanced over the ridgeline, disappeared from sight and were cut to pieces. Communications completely broke down,[113] thanks to enemy shelling, and although both brigadiers made valiant efforts to regain control of the battle, their task was hopeless.[114] Attempts were made to alleviate the crisis by launching the Independent Force into action, but to no avail in the face of intense German machine gun and artillery fire, unimpeded by any Canadian barrage.[115] Mercifully, the enemy retreated during the night to the eastern side of the Canal du Nord. Of the 2800 men sent in to battle from MacBrien's brigade, half the officers and one-third of the other ranks had been killed or wounded, and losses among Odlum's men were only slightly less severe. The official history properly called the 4th Division attack "fruitless, [though] the [overall] results ... were nonetheless eminently satisfactory," a view shared by General Currie.[116] Subsequent analysis by Watson and his brigadiers highlighted why their attack had gone so badly wrong. Suffice to say these were obvious and already known, starting with the folly of sending men into country devoid of cover and swept by machine gun fire without a proper artillery barrage – or any barrage at all – which had been learned on the Somme, two years earlier. To do it in the questionable hope that a convoy of armoured cars, trucks and horsemen would capture bridgeheads, which a properly planned and supported infantry attack could have achieved in two or three days was unfathomable. Currie and Webber, and even more senior British officers, carried the bulk of the blame, while Watson and Panet, given that the plan they were ordered to execute was fundamentally unsound, were rather less culpable.[117] No amount of "hard pounding" could compensate for faulty judgement.

The attack on the Canal du Nord commencing September 27 was the Canadian Corps' last major operation. The attack plan was complex and daring – some would have used stronger language – but nonetheless a tribute to his confidence both in his formations' fighting prowess and in growing German weakness. Attacking with the 1st Division, the 4th Division was to force the canal proper and then press on to clear nearby Bourlon Wood. Although the 11th and 12th Brigades encountered stubborn resistance, and had to cope with a flank exposed by the failure of a British division to keep pace, most objectives had been gained by day's end, and the Germans were falling back. On September 29, both divisions resumed their attacks but made only modest gains. The following day, a failed barrage and stiff German resistance further hampered the 4th Division's progress – "a hellish day," Watson noted in his diary. Only a week earlier, he had been in London on leave.[118] During the afternoon of September 30, Currie assembled his commanders at Watson's headquarters and cobbled together one last thrust by elements of all four divisions that would be launched the following day, but it achieved next to nothing and left the infantry utterly exhausted. Within 24 hours, General Horne wisely called off the operation.[119] While Currie's more ambitious objectives had not been gained, the battle was still a great success. Given difficult tasks and faced with surprisingly stubborn resistance, Currie thought Watson and his 4th Division had fared well.[120]

With the end of the war now in sight, Watson had no desire "to needlessly sacrifice men."[121] German resistance was dwindling, and many days involved more liberating than stiff fighting. On the 20th, he and his men entered Denain whose civilian population "[was] just crazy with joy." The mayor made a fine address, and presented him with a large *tricolor.* Then Watson replied with his own speech in French, much to the delight of the locals.[122] On the October 27, the Prince of Wales along with Generals Currie and Watson and their staffs attended a high mass of thanksgiving in Denain cathedral. Famously recorded in a photograph, Watson looked worn but was beaming, while even Currie had the makings of a smile on his face.[123]

On November 6, the 4th Division was finally relieved – its war was over. The following day, neither Watson nor any of his officers were present for the official ceremony organized by General Horne to mark Valenciennes' liberation, carried out by the 4th Division five days earlier. Currie attended "a very frosty affair" only because Horne ordered him to, accompanied by a handful of his troops who occupied the tail end of the march past.[124]

On the morning of November 11, Watson was telephoned by Corps Headquarters with confirmation of the armistice. "We got the bands of the 10th Brigade together ... and at 11 exactly the mayor raised the French flag and the bands played the 'Marseillaise'. It was a most inspiring sight that after four years of war, we had gained complete victory over the enemy."[125] Watson finally returned to Quebec City late on the evening of July 1, 1919, to be met by Mary and his three daughters. His last diary entry read simply:

> So, after nearly five years of active service, I have returned safe & secure home again, and after what terrible experiences and what fearful hardships and sufferings.[126]

What verdict can we offer on the military career of Major-General David Watson? As a player of wartime politics, he had no equal in the Corps. As a military commander, Stephen Harris called him one of Hughes' "gems" and Desmond Morton one of the Corps' "ablest senior officers." In contrast, Tim Cook dismissed him as "one of the weakest, if longest serving, Canadian divisional commanders of the war."[127]

By 1917-1918, Watson served in a very good army. The superb talent at Corps level made good officers look superior and ordinary ones look good. Beyond that, he was particularly blessed with very able subordinates. His first GSO1, the brilliant and arrogant William Ironside, supposedly so dominated him as to effectively run the division. Ironside lamented years later that he wished the 4th Division "had had a better man than old David Watson [who] couldn't stand the responsibility of command."[128] But as an inexperienced divisional commander – and Watson was very inexperienced in 1916 and 1917 – he would surely have been held to account for not frequently deferring to him. Edouard Panet was also a very able staff officer and an easier personality, too, and Watson was a more mature and presumably more self-confident leader by 1918. Unfortunately for historians, Panet left no opinions on Watson's merits as a general, but as with Ironside, they seemed to work well together. Watson also benefited from excellent brigadiers. In January 1918, Currie named Odlum and MacBrien among the five brigadiers he deemed fit for divisional commands.[129] Hilliam gave Watson good service, and after his transfer from staff work, Hayter gave fine service. There was also quality among the battalion commanders, including the likes of James Clark, Eric Harbottle, Herbert Dawson, Lionel Page, James O'Donahoe, Reginald Davies, Ralph Webb, Herbert Keegan, Robert Palmer, Alfred Carey, Guy Kirkpatrick, James Kirkaldy, Kenneth Perry, Fred Lister and Layton Ralston. In other words, Watson was provided with a good team whose abilities he drew on and generally did not impede. The handful of disenchanted officers is more than matched by the many who continued to serve willingly under Watson's command. Watson was loyal to his men, at least if they were loyal to him, frequently taking an interest in their personal welfare and the promotion of their careers.[130] Morale in the senior ranks of his division seems to have been as high as elsewhere in the Corps. He certainly never had to confront a near mutiny of brigadiers fed up with his authoritarian tendencies as Lipsett had come close to facing.[131] As for the soldiery, they may well have grumbled, but it's a rare commander who avoids that, and there's no evidence that morale among the rank and file wasn't good.[132] Even Ironside, Watson's premier critic, considered the 4th "one of the best [divisions] I have ever seen in any campaign."[133]

As a military mind, David Watson was no Arthur Currie, but neither were any of the other Canadian-born divisional commanders. Watson was an "amateur warrior" in the purest Canadian sense – one whose only military background before August 1914 had been the militia, which is to say little military background at all. He turned out to be very brave, with a willingness (and capability) to learn, and displayed competent if never brilliant insights on the battlefield. Despite his inner flaws – occasions of indecisiveness and poor judgement combined with a propensity to take risks and screaming ambition – he got better. In sum, Watson was neither incompetent nor accomplished. Rather his generalship is best described as competent – good enough for the job and no more. Such generals can – and did – win wars.

---

1 Paul Rutherford, *A Victorian Authority: The Daily Press in Late Nineteenth-Century Canada* (Toronto: University of Toronto Press, 1982).

2 Mark Moss, *Manliness and Militarism: Educating Young Boys in Ontario for War* (Toronto: Oxford University Press, 2001).

3 Ronald Haycock, *Sam Hughes: The Public Career of a Controversial Canadian, 1885-1916* (Waterloo, ON: Wilfrid Laurier University/National Museums of Canada, 1986), p.184. Ibid., pp.183-187.

4 Watson's political and personal friendship with Hughes at the outset of the war is well established, mentioned by both Haycock and Morton. Had Hughes not decided to appoint him to command the 2nd Battalion, which was raised from militia units in Eastern Ontario, Watson might have been out of luck. The Anglophone population of his native Quebec City area was never sufficient to support a battalion.

5 Tim Cook, *At the Sharp End: Canadians Fighting the Great War, Vol. I: 1914-1916* (Toronto: Viking, 2007), pp.153-154 and Nathan Greenfield, *Baptism of Fire: The Second Battle of Ypres and the Forging of Canada, April 1915* (Toronto: Harper Collins, 2007), p.258. See also GWL Nicholson, *Canadian Expeditionary Force, 1914-1919: The Official History of the Canadian Army in the First World War* (Ottawa: Queen's Printer, 1964), especially pp.71-80. On Watson's performance, see Desmond Morton, *A Peculiar Kind of Politics: Canada's Overseas Ministry in the First World War* (Toronto: University of Toronto Press, 1982), p.41.

6 Watson Papers, MG 30 E 69, Library and Archives Canada (LAC), Diary, reel M-10, 24 Apr 1915.

7 Greenfield, p. 260. Watson Papers, Diary, 26 Apr 1915.

8 Morton, Chaps. 2-4. Watson's meetings and correspondence with Aitken were frequent, e.g. Watson Papers, Diary, 2, 3, 11 and 27 Aug 1915.

9 Nicholson, pp.137-145.

10 Cook, *Sharp End*, pp.340-341.

11 Beaverbrook Papers, LAC, MG 27 II G 1, Reel A-1764, Aitken to Hughes, 19 Jan 1916. Haycock, p.281.

12 Watson Papers, Diary, 1 Mar 1916. Stephen Harris, *Canadian Brass: The Making of a Professional Army, 1860-1939* (Toronto: University of Toronto Press, 1988), p.118. Burstall was a pre-war regular which alone would have doomed him in the Minister's eyes.

13 Morton, pp.60 and 68. On Currie's respect for Watson, Beaverbrook Papers, A-1764, Aitken to Hughes, 16 Oct 1915. Watson's view of the Ross was that he had "no report that the men are without confidence in the Ross Rifle," pointing out correctly that all weapons had jammed in the St. Eloi mud. Two of his four battalion commanders disagreed, however. Ibid., A-1765, *Opinions of Brigade and Battalion Commanders on the Ross Rifle*, undated [early May 1916].

14 Ibid., A-1764, Aitken to Hughes, 20 Jan 1916. Ibid., A-1765, Watson to Aitken, 10 and 26 Mar 1916.

15 Watson Papers, Diary, 4 Apr 1916. Ibid., 5 April 1916.

16 Haycock, pp.292-294 and Morton, pp.63-69. Aitken later revealed that he'd informed the British that the post of Inspector General of Canadian forces in England was to be a "nominal one", information one suspects Watson's prying questions must have extracted during their behind-the-scenes discussions. Beaverbrook Papers, A-1764, 10 May 1916. The shrewd Aitken later confided to Hughes many felt that "W[atson] didn't even want to be in England and it was just a stepping stone to France" all along. Ibid.

17 Watson Papers, Diary, 28 Apr 1916. Ibid., pp. 23, 25, 26 and 27, Apr 1916.

18 Department of Militia and Defence, LAC, RG 9, III A 1, v. 44, file 8-5-8E, Watson to Currie, 12 May 1916. See also Beaverbrook Papers, Reel A-1765, Watson to Aitken, 8 Aug [1916].

19 Ibid., A-1764, Aitken to Hughes, 4 May 1916.

20 Ibid.

21 Watson Papers, Diary, 4 May 1916. Ironside agreed wholeheartedly. Beaverbrook Papers, A-1765, Ironside to Aitken, 8 May 1916.

22 Watson Papers, Diary, 8 May 1916.

23 Ibid., 30 Jun 1916.

24 Ibid., 30 Jun and 8 and 10 Jul 1916. Rexford only agreed to step down if he was allowed to revert to second-in-command, and this was agreed to.

25 Ibid., 30 Jun 1916. Ibid., 1 Jun 1916.

26 Beaverbrook Papers, A-1765, Watson to Carson, 8 May 1916.

27 Ibid., Aitken to Watson, 5 Jul 1916.

28 Max Aitken, no amateur when it came to the subtle arts of flattery, ingratiation and persuasion, not to mention self-promotion, had once commented favourably on Watson's "great

diplomacy and excellent judgement" in securing the resignation of a deficient commanding officer in his brigade. Ibid., A-1764, Aitken to Hughes, 16 Oct 1915.

29 Ibid., A-1765, Watson to Aitken, 10 Aug 1916. Ibid., A-1764, Watson to Aitken, 8 August [1915]. Ibid., Aitken to Ironside, 9 May 1916.

30 Watson Papers, Diary, 6 Jun 1916.

31 He at least trusted him with money, for Aitken was still making stock purchases for Watson in December 1916. Beaverbrook Papers, A-1765, Watson to Aitken, 1 Dec 1916. Watson also continued to employ Aitken's son, Captain Alan Aitken, as a GSO 3 at his headquarters. Andrew Godefroy, "The Fourth Canadian Division: 'Trenches Should Never Be Saved,'" in Geoffrey Hayes *et al.*, eds., *Vimy Ridge: A Canadian Reassessment* (Waterloo, ON: Wilfrid Laurier University Press, 2007), p.214.

32 Harris, p.113.

33 Watson Papers, Diary, 16 Jun 1916. Ibid., 17 Jun 1916. On the British view, see Beaverbook Papers, A-1765, Wigham to 4th Division HQ, 7 Jun 1916. As for reinforcements from the 4th Division, these commenced with a first draft of 2500 men on Jun 8 and a second of 1075 nine days later. Watson Papers, Diary, 8 and 9 Jun 1916. By the end, around 8000 men were taken. Andrew Godefroy, "Fourth Division," p.214. Aitken tried to console him by suggesting that, having made his sacrifices, Canadian authorities and the War Office would likely look even more favourably on the 4th Division's manpower needs. Watson Papers, Diary, 10 Jun 1916.

34 Ibid., 12 May 1916.

35 Beaverbrook Papers, A-1764, Hughes to Aitken, 19 May 1916. Leopold Guy Frances Maynard, Lord Brooke had been serving with the Canadian militia in 1914, and from Hughes' perspective was the right kind of Briton, an aristocrat and an admirer of Hughes.

36 Ibid., Hughes to Aitken, 24 Jun 1916.

37 Watson Papers, Diary, 10 May 1916. Ibid., 8 May 1916. On Watson's falling out with Loomis, see ibid., 13 May 1916.

38 It has been argued that Sir Sam Hughes delayed the deployment of the division to France to give himself more time to get his way with the brigadier appointments. Harris, p.117. Once the division left England, officer appointments would have been the sole prerogative of the Corps commander, Lieutenant-General Byng, for that was one of the terms he extracted in accepting the post in late May, as Hughes would have been well aware. Patrick Brennan, "The Other Battle: Imperialist versus Nationalist Sympathies within the Officer Corps of the Canadian Expeditionary Force, 1914-1919," in Douglas Francis and Phillip Buckner, eds., *Rediscovering the British World* (Calgary: University of Calgary Press, 2005), pp.253-254.

39 Others would have included Brigadiers Griesbach and Ketchen and General Turner.

40 Watson Papers, Diary, 15 Nov 1916.

41 Morton, p.119. Watson Papers, Diary, 22 Nov, 4 and 7 Dec 1916 and 10 Jan 1917. Predictably, Brigadier St. Pierre Hughes argued Watson was driven by political animus and the need to cover his own incompetence – Byng's decision clearly supports Watson's version.

42 Beaverbrook Papers, A-1765, 10 Aug 1916.

43 Nicholson, pp.188-189.

44 Watson Papers, Diary, 22 Sep 1916. Ibid., 18 Sep 1916.

45 Ibid., 23 Sep 1916.

46 Ibid., 28 Sep 1916. His detailed diary records many more visits to Odlum's headquarters than either MacBrien's or Hughes's. More than just colleagues, a close friendship had quickly blossomed between Odlum and Watson. His relationship with MacBrien was more business-like. There was no relationship at all with Hughes, just suspicion and contempt.

47 The best account of the 4th Division on the Somme is given in Cook, *Sharp End*, Chap. 38.

48 Nicholson, p.191.

49 Watson Papers, Diary, 2 Nov 1916.

50 Nicholson, p.192.

51 Ibid., p.193.

52 Ibid., pp.194-197.

53 Patrick Brennan and Thomas Leppard, "How the Lessons Were Learned: Senior Commanders and the Moulding of the Canadian Corps after the Somme," in Yves Tremblay, ed., *Canadian Military History Since the 17th Century: Proceedings of the Canadian Military History Conference, Ottawa, 5-9 May 2000* (Ottawa: Department of National Defence, 2001), pp.138-139 and Bill Rawling, *Surviving Trench Warfare: Technology and the Canadian Corps, 1914-1918* (Toronto: University of Toronto Press, 1992), Chap. 4.

54 LAC, RG 9 III C 1, v. 3954, folder 93, file 3, 3 Nov 1916.

55 Ibid., v. 3964, folder 99, file 4, Watson report on re-organization in the Canadian Corps. 20 Dec 1916. Ibid., Lipsett report on re-organization in the Canadian Corps.

56 Watson Papers, Diary, 26 May 1915.

57 Ibid., 22 Feb 1917.

58 LAC, RG 9 III C 1, v. 3858, folder 83, file 4, Odlum report, 17 Mar 1917. Tim Cook, "'A Proper Slaughter': The March 1917 Gas Raid at Vimy Ridge", *Canadian Military History*, 8, 2 (spring 1999): 14-19.

59 Watson Papers, Diary, 1 Mar 1917.

60 Ibid., 2 Mar 1917. Watson seems to have felt genuine grief over the heavy losses among his officers and men.

61 Brian Bond, "Ironside", in *Churchill's Generals*, John Keegan, ed. (London: Warner Books, 1991), p. 18. Cook, "Proper Slaughter," pp.20-21. See also Tim Cook, *No Place To Run: The Canadian Corps and Gas Warfare in the First World War* (Vancouver: UBC Press, 1999), 97-98 and fns 25 and 36, 256.

62 Beaverbrook Papers, A-1765, Aitken to Borden (draft), 10 Sep 1916. While acknowledging his "recognized reputation" in the field, the phrase Aitken used was "his indecision invariably leads to confusion." He dropped it from the letter sent to the Prime Minister.

63 Watson Papers, Diary, 22 Feb 1917. Ibid., 6 Feb 1917.

64 Ibid., 18 Feb 1917. Unfortunately Watson's estimate of the minimum artillery barrage required was either too low to start with or subsequently watered down. Ibid., 14 Feb 1917.

65 LAC, RG 9 D III, v. 4952, War Diary, 54th Battalion, 1 Mar 1917.

66 Cook, "Proper Slaughter," p.21.

67 RG 9 III C 1, file 16, Watson to Corps, re: platoon at Vimy, 9 May 1917.

68 Godefroy, "Fourth Division," p.222. Ibid., pp.217-220.

69 Nicholson, p.259. Godefroy, "Fourth Division," p.221 and n. 26, p.224.

70 Richard Turner Papers, LAC, MG 30 E 46, v. 11, file 80, Manley-Sims to Turner, 28 Apr 1917.

71 The replacement battalion for the 73rd was Nova Scotia's 85th, commanded by the Prime Minister's cousin, Lieutenant-Colonel Arthur Borden.

72 Watson Papers, Diary, 8, 10 and 20 Mar 16 and 18 Apr 1917.

73 Patrick Brennan, "Julian Byng and Leadership in the Canadian Corps," in Geoffrey Hayes *et al.*, eds., *Vimy Ridge: A Canadian Reassessment* (Waterloo, ON: Wilfrid Laurier University Press, 2007).

74 Watson Papers, Diary, 8 Jun 1917.

75 Ibid., 27 Nov 1916.

76 Watson had been suggested for the job, but Perley, Borden's representative in England, considered him "junior and still gaining much experience by remaining at the front." Morton, p.97.

77 Currie's only rival (within the Corps) was Major-General Louis Lipsett, who had been with the CEF from the outset of the war, rising from battalion command, but was also a British regular and therefore unacceptable politically. Looming in the background, of course, was Major-General Garnet Hughes, the "Mad Mullah's" son, but all agreed his selection was unthinkable.

78 Watson Diary, 5 Feb 1917. Ibid., 4 Feb 1917.

79 When Ironside matter-of-factly related that Byng felt his replacement would have to be "Currie or an Imperial", Watson rejected the latter option out of hand. Ibid., 8 Jun 1917.

80 Morton, pp.122-123 and AMJ Hyatt, *General Sir Arthur Currie: A Military Biography* (Toronto: University of Toronto Press, 1987), p.74.

81 Beaverbrook Papers, A-1764, Hughes to Aitken, 17 Aug 1917.

82 A good example of this was William Griesbach. Promoted by Byng to command the 1st Brigade in February 1917, he continued in that capacity until war's end, but was not part of Currie's inner circle and was never considered for promotion. Griesbach was known to have continued corresponding with Hughes after his removal from the cabinet.

83 Nicholson, pp.284-292.

84 Ibid., pp.292-297 and Geoffrey Jackson, "'Anything but Lovely': The Canadian Corps at Lens in the Summer of 1917," *Canadian Military History*, 17, 1 (winter 2008).

85 Ibid., pp.17-19. RG 9 III C 1, v. 3851, folder 63, file 9, "10th Canadian Infantry Brigade Operations, 29 August 1917."

86 "A truly maginificent performance", General Currie noted. Arthur Currie Papers, LAC, MG 30 E 100, v. 43, file 194, Personal Diary, 26 Oct 1917. Watson Papers, Diary, 28 and 29 Oct 1917.

87 Ibid., p.2 and 6 Nov 1917.

88 Currie Papers, v. 2, General Correspondence file, 1915-18, M-R, Currie to Perley, 10 Nov 1917.

89 Watson Papers, Diary, 2 Dec 1917.

90 Ibid., 17 Dec 1917. Ibid., 19 Dec 1917.

91 CEF Personnel Records, LAC, RG 150, Acc. 1992-93/166, E deB Panet file.

92 Watson's diary entries are often quite revealing of his emotions – there is just one brief entry mentioning "good-byes from Ironside." Watson Papers, Diary, 26 Dec 1917. The Corps policy called for the "Canadianization" of all senior positions, meaning that as soon as a competent Canadian had been trained, he took over permanently. Simply put, Panet's time had come.

93 John Warden Papers, LAC, MG 30 E 192, Diary – 1918, 2 Jan 1918. Ibid., 1 Jan 1918. Watson's diary does not mention either the incident or its buildup, and neither do Currie's diary nor Odlum's papers.

94 Watson Papers, Diary, 27 Mar 1918.

95 Brennan and Leppard, pp.139-141.

96 Watson Papers, Diary, various entries May 13-Jul 12, 1918.

97 Ibid.

98 Ibid., 1, 3, 4, 6-7 Aug 1918.

99 Nicholson, pp.394-398 and pp.404-419.

100 Watson Papers, Diary, 14 Aug 1916. Ibid., 16 Aug 1918.

101 Currie Papers, v. 43, personal diary, 1914-1919, 24 Aug 1918.

102 Watson Papers, Diary, 29 Aug 1918.

103 Daniel Dancocks, *Spearhead to Victory: Canada and the Great War* (Edmonton: Hurtig, 1987), p.108.

104 Nicholson, p.434.

105 LAC, RG 9 III D 3, v. 4905, War Diary, 11th Brigade, v. 4905, General Staff Report for period 17 August to 22 August 1918, authored by Odlum 22 Aug 1918. Also ibid., v. 4861, War Diary, 4th Division, *Report on Scarpe Operations/Second Battle of Arras*, 2. See also ibid., v. 4909,

War Diary, 12th Brigade, *Report on Scarpe Operation – Capture of Drocourt-Queant Line 2nd September 1918*, pp.1-2. Ibid., v. 4861, War Diary, 4th Division, 30 Aug 1918.

106 Currie Papers, v. 43, file 195, General Notes, undated [late Aug 1918]. RG 9 III D 3, v. 4861, War Diary, 4th Division, *Narrative of Operations of 4th Division from 26th August to 4th September 1918*, pp.6-7.

107 The Independent Force was a mixed unit of armoured cars, truck-mounted mortars, cyclists, British and Canadian cavalry, and the two Canadian Motor Machine Gun Brigades.

108 Watson Papers, Diary, notes section, 8 Aug 1918.

109 Turner Papers, v. 8, file 51, Currie to Turner, 29 Mar 1918.

110 LAC, RG 9 III D 3, v. 4987, War Diary, 2nd Motor Machine Gun Brigade, Canadian Independent Force report. Ibid., v. 4861, War Diary, 4th Division, *Report on Scarpe Operations/Second Battle of Arras*, p.3.

111 Ibid., v. 4905, War Diary, 11th Brigade, Operation Order No. 150, 1 Sep 1918.

112 Ibid., Appendix 2, *Narrative of Operations carried out between Sept. 2nd and Sept. 5th*, 1918 ..., p.2.

113 The radio set at 11th Brigade headquarters was knocked out by a German shell within 5 minutes of the start of the attack and thereafter the land lines worked only intermittently. This effectively put Watson and Panet in the dark. See also Watson Papers, Diary, 2 Sep 1918.

114 LAC, RG 9 III D 3, v. 4905, War Diary, 11th Brigade, Appendix 2, *Narrative of Operations carried out between Sept. 2nd and Sept. 5th, 1918* ..., pp.2-7. Ibid., v. 4861, War Diary, 4th Division War, *Report on Scarpe Operations/Second Battle of Arras,* Article III, p.6.

115 Nicholson, p.438.

116 Ibid., p.438. Currie Papers, v. 43, Personal Diary, 1914-1919, 2 Sep 1918.

117 Dancocks lays most of the blame on Watson's poor generalship. Dancocks, pp.115 and 117ff.

118 Watson Papers, Diary, 29 Sep 1918. During the course of the war, Watson only took four two-week leaves, turning down many other opportunities.

119 Nicholson, pp.445-453.

120 Watson Papers, Diary, 3 Oct 1918.

121 Ibid., p.26, Oct 1918.

122 Ibid., p.20, Oct 1918. Denain had been "captured" by 4th Division patrols the previous day, culminating the longest single-day advance of the war by the Corps, some 12000 yards (nearly 7 miles). It is a sad reflection of the legacy of trench warfare fighting that advances were still being measured in yards.

123 Ibid., p.27, Oct 1918. Twelve days earlier, Watson had been offered the postwar command of the Montreal Military District at a salary of $6000. Ibid., 15 Oct 1918. Many such offers were made to senior officers in the last months of the war, the Borden government's intention

seemingly having been to honour (and reward) these men for services rendered to the country (and costly time away from businesses and professions). Odlum, for example, had been earlier offered the Customs inspectors post in Vancouver.

124 Currie Papers, v. 43, Personal Diary, 1914-19, 7 Nov 1918. Ibid., 5 Nov 1918 and Watson Papers, Diary, 7 Nov 1918. There was noticeable friction between Currie and the Canadian Corps and Horne and his First Army staff in the last weeks of the war, driven more by questions of Canadian professional pride and what constituted appropriate recognition of the Corps' achievements than nascent Canadian "nationalism". Brennan, "Other Battle," pp.255-257.

125 Watson Papers, Diary, 11 Nov 1918.

126 Ibid.

127 Harris, p.119; Morton, p.91; and Cook, *Sharp End*, p.341.

128 William Ironside Papers, IWM 92/40/1 [Imperial War Museum], Ironside to Lindsey, 15 Mar 1946.

129 Turner Papers, v. 8, file 51, Currie memorandum, undated [Jan 1918].

130 In July 1917, Watson supported Odlum's decision to remove Lieutenant-Colonel Valentine Harvey, the commanding officer of the 54th Battalion, after he was court-martialed for being absent from his battalion, along with his second-in-command, when it was in reserve. A month later Watson asked Currie not to have Harvey sent back to Canada in disgrace, as was the practice, but to employ him usefully in England and he was even prepared to see him return to the 4th Division for some special duty. "Harvie [sic] is an exceptionally good officer," Watson stressed, "and the penalty which he has now paid for his mistake, will never be forgotten, I am certain." Currie Papers, v. 2, General Correspondence, 1915-18, S-Z, 31 Aug 1917.

131 Ibid., v.1, General Correspondence, 1915-1918, A-F, Elmsley to Currie, 17 Sep 1918.

132 There were only three executions in the 4th Division, out of 25 in the Corps. Andrew Godefroy, *For Freedom and Honour: The story of the 25 Canadian Volunteers executed in the Great War* (Nepean, ON: CEF Books, 1998). Watson had a pragmatic streak when it came to dealing with potential morale problems. Dismissing the complaints of prominent Protestant clergy that he was too tolerant of Roman Catholic chaplains in some of his battalions, an exasperated (Presbyterian) Watson once blurted out "But I have a lot of Catholic soldiers!" On Watson's sentiments about the stupidity of religious divisions within the army, see Watson Papers, Diary, 16 Sep 1917.

133 Ironside Papers, Ironside to Lindsey, 20 Oct 1955.

CHAPTER 6

Leadership and Innovation

Andrew McNaughton and the Counter-Battery Staff Office

PAUL DICKSON

As the year 1916 closed, few would have suspected that they had planted the seeds of an Allied victory on the western front in 1918. The butcher's bill of the Somme and Verdun convinced many observers, and participants, that the war would not end anytime soon. But in the British and French armies, the hard lessons of the previous years fighting were being identified and applied to organizations and tactics. Experiments with new technology and innovative uses of old technologies were slowly transforming operations. This change was evolutionary; a response to conditions, and was bottom-up, even if some of it was later standardized by and disseminated from the top.

The Canadian Corps was a leading practitioner of this learning practice, a system that today we call an after action review and lessons learned process.[1] Why this was so has been the subject of much debate. As the weight of evidence of the learning that went on in the British and Dominion forces during the Great War has accumulated, the central issue has shifted from trying to dispel the old stereotypes of "lions led by donkeys" or "red-tabbed dopes" to attempts to identify the origins of the innovation and military excellence demonstrated by so many commanders and staff officers.[2] Were there national or professional cultures that predisposed the Canadians and Australians to learn and innovate? That slowed learning in the British army? Was it down to leaders? Or the specific circumstances faced by each organization? What, for instance, was gained from organizational stability and long periods employed on one front? The evolution of the Counter-battery Staff Office (CBO) provides some insights into these questions.

Recent studies characterize the development of counter-battery techniques as central to the successes of the British and Dominion forces during the last two years of the war and as exemplary in reshaping the traditional view that military professionals failed to meet the demands of trench warfare. Rather the evolution of the CBO demonstrates the interplay of innovation, leadership, personality and emerging professionalism that harnessed technology and science to create an all-arms operational method that proved capable of restoring some semblance of mobility to the battlefield. As one author suggests, the CBO was "a reflection of the vibrancy of British experimentation and their determination to find solutions to find solutions to the stalemate on the western front."[3] Central to the development of counter-battery efforts in the Canadian Corps was a future army commander, the first Counter-Battery Staff Officer (CBSO) Lieutenant-Colonel A.G.L. "Andy" McNaughton.

The period from 1914-1916 was rich with lessons, some observed and learned, some not. The artillery's emergence as a war winner was one lesson slowly digested by all combatants, hindered primarily by the limited supply of artillery, shells and decision-makers who could appreciate the changes being wrought by the machine gun, artillery, barbed wire and spade; by 1916, it was becoming evident that those who best maximized the artillery's capabilities would achieve the breakthrough both sides so desperately sought. And it was a question that was intimately tied to the combatant's industrial and technological capacity and capability, their military's tolerance for innovation and change, and, perhaps most important, the emergence of key individuals who embodied the latter and understood the former. Andrew McNaughton was one such individual. The purpose of this chapter is not to suggest he was the first to understand the potential of the tools at his disposal; he was one of a number of individuals in the British Imperial forces who came to this understanding. But McNaughton was able to achieve a degree of success as the CBSO in the Canadian Corps that eluded his counterparts in other corps, and he was able to achieve it much earlier, by April 1917. And it was, in part, his success and the consequent credibility that victories like Vimy bestowed on scientific gunnery, or the "siege gunner fandoodle" like flash and sound ranging, meteorological reports, and surveying, that convinced even the most reluctant to adopt these methods across the British army.[4]

No armies were fully prepared for the artillery-based operations that would characterize and eventually win the war for the western Entente. Despite the hold of images of waves of men mown down by machine guns, artillery was by far the biggest killer of the war, and silencing the enemy's guns became the precondition for success in all operations. The Franco-Prussian War appears to have exerted the greatest influence on artillery organization, doctrine and tactics prior to 1914. It has been argued that the two principal lessons from that conflict were the need to mass the artillery and the importance of diverting the attention of the enemy's artillery, the "artillery duel." These principles were rooted in some basic assumptions, which derived from the technology of the period and the dominance of the field artillery within that branch of the service. The main assumptions were that the artillery could only destroy what it could see and that artillery's role was determined by the infantry plan; despite advances in technology, these assumptions carried the artillery into 1914.[5]

Artillery fire had its limits until 1916. The British army was not capable of mounting effective artillery barrages much less counter-battery operations during 1914-1915.[6] The shortages of heavy artillery are well documented, as are the problems with shells, both in quantity and quality. The British had 504 guns in total in August 1914; when the war ended, they had 6406, including over 2200 heavies.[7] Just as important were the shortages of good maps, sensors, aircraft for observation and a good communication system. The shortfalls that minimized the artillery's effectiveness were exacerbated through 1915-1916 by the general hostility to scientific innovations the resulted from the predominance of the field artillery as the main adjunct to the infantry; siege and garrison artillerymen, who dominated the heavy artillery and were long used to primitive flash-ranging and accounting for meteorological conditions, were viewed as eccentric cousins within the service. Adoption of better organization and practices was also hindered by the lack of an adequate command and control system for the artillery. Even the most senior artillery officers were advisors to the formation commanders; they could not impose their plans across the artillery units much less on the formation commanders whom they advised. That full extent of the weakness of the artillery barrage as a system in the British army was made clear on the first day of the Somme. It had been assumed that the shortages in guns and shells had been made good by 1 July

1916, but it was evident – "in a sad scrawl of broken earth and murdered men" – that they had not.[8]

The search for solutions began before the battle ended, and the establishment of the CBO, headed by a staff officer, exemplified one of the most important conclusions: an affirmation of the corps headquarters as the primary operational unit, particularly with regards to control of the artillery. Equally important, the establishment of the CBO represented the creation of a weapons system. As such the creation of an effective counter-battery staff office required the evolution of several complementary technologies, systems and practices, many at rudimentary stages, or non-existent, prior to 1914. The development of these weapons' systems was certainly the precondition for successful counter-battery efforts and a critical factor in victory.[9]

Information, intelligence and communications were the pillars upon which the CBO would be built.[10] If the gunners were to shoot at targets they could not see, they had to find out where they were. Shooting from the map, without ranging or registration, required information and a level of detail in maps – an understanding of the relative positions between guns and targets – that did not exist in 1914. In 1914, the British army's Royal Engineers were responsible for survey work and GHQ had the only topographical section. Through 1915, 32 million maps were produced. By 1917, there was one survey section per corps and surveyors accurate and increasingly large-scale maps became keys to developing effective artillery and infantry tactics. In 1918, specialized surveyors were provided to divisions. In the Canadian Corps, the Topographical Section was a sub-section of the Canadian Corps Survey Section under the Intelligence Branch. This Topographical Section supplied the necessary information on maps to all formations in the corps, a task made simpler, though not simple, by the relatively static conditions of the western front.[11] The Canadian Corps experimented with specialization even further, forming Engineering Battalions attached to divisions. In the corps commanders General Sir Arthur Currie's view, these battalions made one of the most significant contributions to victory.[12]

Accurate and properly scaled maps, with the location of the artillery, still required information on the targets and the ever-changing landscape of the western front defensive systems. Prior to the war, neither the French nor the British armies contemplated an organization that would gather, assess and distribute information for the dedicated use of the artillery. Soon, the French created the "Service de Renseignment de l'Artillerie" and the British the "Reconnaissance Service." The primary source for this information from 1914 on was accurate and current aerial photographs, a source of information that depended on control of the air. The link between air superiority and the flow of intelligence and surveillance was evident early in the war; gradually improvements in communications and photographic technology and communications procedures – much of it driven by the needs of the counter-battery staffs through 1916 – provided sources of information that were indispensable to the effectiveness of the counter-battery efforts.[13] The effectiveness of the Canadian and British counter-battery techniques would not have been fully realized without the air supremacy achieved by 1918, a contribution that should not be overlooked.

Communications was another problem that had to be solved before an effective counter-battery system was realized. In general, communication problems bedevilled combatants on both sides, and were, in the words of one particularly good assessment, "a problem that was almost insoluble."[14] With regards to command and control of the troops during operations this was true. But the problem was less insoluble for the artillery, and for the heavy artillery on static fronts it proved difficult, but not impossible. Technology provided part of the solution, but in the end it was a numbers game and redundancy became the best guarantee of successful communication: quantity had a quality all its own. By 1915, communications were laddered; during the battles on the Somme, communications was conducted by telephone, pigeon, visual signalling with lamps and flags, and runners.[15] Wireless communication was used at Vimy, but the state of the technology led Currie to conclude that wireless was at best subsidiary to other means of communication.[16]

McNaughton identified "a flexible system of command" as one of the keys to effective counter-battery work, a conclusion reached by others as they assessed the CBO at the end of the war.[17] There was a correlation between the artillery's importance and the evolution of their positions as ones that were strictly advisory to ones of executive command, although some ambiguity remained to the end of the war. In 1914, the senior gunner at the corps level was the Brigadier-General,

Royal Artillery (BG, RA), who acted in an advisory role and had no control over the heavies of the Royal Garrison Artillery, controlled at the army level. Given that much of the science at the core of the artillery's dominance came from this branch, it was a potentially disastrous divide on a number of levels.[18] Experience proved just how disastrous and in the aftermath of the Battle of Loos in the fall of 1915, the BG, RA was elevated to General Officer Commanding, Royal Artillery, an improvement, but still a small step. In March 1916, just how small was illustrated when a Commander, Heavy Artillery (CHA) was created as another corps level position intended to provide a bridge between army control of the heavy artillery and the reality that the corps headquarters controlled battles.[19]

The failure to destroy the German artillery and the evident utility of heavy artillery during the Somme battles in 1916 proved critical to changing the artillery's command and control structure. Through the winter of 1916-17, the British institutionalized many of the lessons of the Somme. GHQ issued a number of pamphlets to disseminate the lessons of the Somme.[20] The corps headquarters was recognized as the principal operational headquarters; there was a concurrent recognition that, in the static conditions of the western front, artillery was the key to the breaking the deadlock. Consequently, GHQ attempted to standardize establishments, and thus innovations, at corps headquarters across the British army. At the beginning of 1917, GHQ ordered the expansion of the corps' headquarters. GHQ directed that the GOC, RA of the corps would now control all artillery assets, although the corps commander would ultimately determine their use. Among the notable additions to the headquarters were a dedicated Intelligence GSO2 and a Corps Machine Gun Officer. The elevation of intelligence and the centralizing of assets under corps control also underlay the addition to the GOC, RA's staff of a lieutenant-colonel as a Counter-Battery Staff Officer, with a Staff Captain and an aide-de-camp who acted as a Reconnaissance Officer. Tellingly, the intelligence function, which had controlled by GHQ, was centralized in the heavy artillery under the purview of the CBSO.[21]

The origins of a dedicated CBO organization throughout the British army dates from these initiatives in the winter of 1916-17, although elements of the system were evident in the Canadian and other corps before it was brought together and given a formal establishment. As the initiating order suggested it was "owing to the importance of an efficient counter-battery system, as demonstrated in the Somme battle" and in recognition that "'good organization' was the main contributing factor to counter-battery efficiency."[22] The capability if not the organization was recognizable in British V Corps, under the tutelage of Lieutenant-Colonel A.G. Haig. He might rightly be called the father of the counter-battery staff office. He is credited by at least one author as having been the inspiration for GHQ's decision to create counter-battery staffs across the British army.[23] There is evidence that British III Corps established the first official CBO in November 1916; when in January 1917 GHQ ordered all corps headquarters to establish one, British VIII Corps was first off the mark.[24] Still, seeking to uncover who was first obscures the more important point about their establishment: the culture of innovation and the sharing of lessons that characterized the British and Dominion approach to operations in the aftermath of the Somme. And tracking the origins doesn't explain why some corps' CBO evolved faster than others.

The Canadian Corps was slower off the mark, although clearly their CBO was distinct and quickly among the most proficient. McNaughton observed that by 1918 the Canadian Corps' CBSO "drill and routine...was substantially different from what other people were doing in their Counter-Battery Offices."[25] When the Americans arrived in late in the war, after a review of practices across the French and British armies, they patterned their Counter-Battery Office after "that of the British, or more strictly, that of the 1st British and Canadian Corps, which were taken as models."[26] From an operational perspective, the Canadian Corps CBO surpassed the majority of its counterparts in effectiveness, both in counter-battery practice and perhaps in the systematic collection and dissemination of intelligence. Why was this so? McNaughton was the catalyst for the CBO's effectiveness. He took a systematic approach to developing his CBO, was curious about new ideas, willing to experiment and forcefully led the CBO where he wanted it to go. The pillars of his success were a highly organized system of intelligence, and the concentration of control of artillery resources he was able to achieve. The clarity of his ideas helped to smooth out some of the ambiguity surrounding his new post. And he benefited from both the organizational climate of the Canadian Corps and the consequent confluence of like-minded individuals willing to let him lead as long as success followed.

The creation of an organizational climate that encouraged learning and innovation in the Canadian Corps is well established. An organization's climate is short-term nature, which distinguishes it from culture. It has been suggested that climate is the more "malleable and responsive" aspect of a deeper culture, susceptible to "immediate pressures and policy guidance." Leaders are crucial in establishing the climate through a transparent system of rewards and punishment, by the values and priorities they emphasize and how they define and measure progress. It is a useful concept because it encompasses how the organization thinks, not just what it thinks; and it reflects how an organization responds to its circumstances, in short how it learns.[27] Charting the process of how an organization evolves and is encouraged to think and learn is challenging, but it is clear that the organizational climate that evolved in the Canadian Corps from 1916 to 1918 was one in which there was an openness to new knowledge and ideas.[28]

Sir Sam Hughes established some of the key elements of the climate of the Canadian Corps; he encouraged, demanded even, breaking with regular British Army conventions. Casting aside conventions did not make the corps unconventional or even innovative, but it set a tone.[29] The first General Officer Commanding of 1st Canadian Division, Lieutenant-General Edwin Alderson, a British officer, also demonstrated a willingness to make his own decisions. He cultivated that approach among his officer corps. General Sir Julien Byng built on that, creating, in the words of one study, "a conducive environment for passing tactical lessons up the chain of command" General Sir Arthur Currie formalized this process and the culture, rewarding innovation and experimentation. The same study aptly summarizes that "what resulted was nothing less than a system that provided continuity of purpose, intentional learning and institutional memory."[30]

A key principle underpinning the innovations of the corps was that merit, not political contacts or nationality, was the primary consideration for promotion. From 1915, as the Corps expanded, Currie fought to install merit as the sole measure of promotion, noting that it was not a question of "whether a man was Canadian or otherwise, it is one of the best man for the job."[31] One by-product of the elevation of professional measures over national was that the Canadian Corps, an iconic Canadian symbol, was also viewed by some officers, after the war, as a model of imperial cooperation, and promise.[32] If it seems to run counter to the spirit of the corps to examine the contributions of British staff officers loaned to the Canadians, it is still useful to note that the practice of counter-battery work suggests that these officers may have flourished in the corps in a way they couldn't in comparable British formations. As noted below, many of the innovators in counter-battery practices, while holding appointments in British formations, first found operational expression of their ideas through McNaughton's applications. Nationality was no bar within the corps. McNaughton took his adjutant Lennox Napier, another British staff officer on loan, with him and credited him with helping to create the CBO.[33] And although they came to loath each other, the overall quality of the British staff officers working with the McNaughton and the artillery is represented by the future Lord Alanbrooke.[34]

So, when McNaughton was chosen to establish the CBO in the Canadian Corps he was fortunate to begin in an environment and in circumstances where, supported by capable and innovative people, he would be allowed to make his own mistakes. McNaughton seized this opportunity. He did not share any of the skepticism with regards to the potential of the new tools that science was making available for sensing that some studies suggest hindered others from the artillery community. He was an engineer by training, and a scientist at heart. Neither, and perhaps more important, did he experience any of the resistance met by some of his counterparts in the artillery. The Canadian gunner community was dominated by the field artillery; sprinkled among CEF that went overseas in 1914, were the gunners who would shape the culture of the corps and the artillery: McNaughton, General Sir Arthur Currie, "Dinky" Morrison, and "Harry" Crerar.[35] The Canadian Corps was also fortunate in that in General Sir Julian Byng and General Sir Arthur Currie, a gunner, they had commanders who recognized the value of the artillery. Currie's well-known report on French innovations during the Verdun counter-offensives stressed the importance of counter-battery fire; he concluded that locating and destroying the enemy batteries should be the main occupation of the artillery before any offensive. It was, in his view, an "artillery war."[36]

Once charged with establishing the CBO, McNaughton visited French and British headquarters, gathering the lessons of the Somme and Verdun. As McNaughton looked across the allied armies for best practices, the Brigadier-General, General Staff at corps headquarters, P.de B. Radcliffe, a British officer who McNaughton credited with much of the efficiency at the headquarters, directed him towards Lieutenant-Colonel Haig at British V Corps. By his own account, McNaughton was dissatisfied until he found in Haig a mentor who matched his own curiosity and dedication to the application of scientific gunnery. Interviewed decades later, McNaughton observed that the "British Army and Canadian Corps owe very deep gratitude to Colonel Haig, not only for his methods, but for the help he gave us in organization."[37] Here was a true meeting of the minds. McNaughton's biographer credited Haig for bringing to McNaughton's attention the most recent developments in sound-ranging, flash-spotting and aerial reconnaissance, and how these were being used to locate the enemy's guns.

The importance of how eagerly McNaughton recognized and seized on the multiple technological developments was evident in the case of sound-ranging, notable both for what is illustrates about the range of methods that McNaughton would use to create one of the premier information gathering systems of the western front as well as for his approach to incorporating new equipment and technologies into counter-battery practice. The official historian of the Royal Regiment of Canadian Artillery compared the reception McNaughton gave the "small group of scientists" from England with that by other CBSO and gunners across the British army. Using the delicate sound-ranging instruments – an oscillograph that recorded sound on film – in trenches was, in the official historians characterization, considered by most to be "madness." Not so in the Canadian Corps, and the innovation added further to the CBO's effectiveness, providing it for a time with a sensor no other corps had. The leading proponent of sound-ranging was a Nobel prize winner for his work in physics, Lieutenant Lawrence Bragg, who commanded the first sound-ranging section. He and McNaughton would become fast friends.

But McNaughton's gift was also his ability to recognize the limits of technology. His experience with sound-ranging, among other developments, prompted him to conclude that a wide variety of sensors was critical to counter-battery work in particular, and the gathering of intelligence in general. In a lecture given to the Staff College, Camberly after the war and later published in *Canadian Defence Quarterly*, he listed no less than twelve sources of "artillery intelligence"; he rated that information gathered by aircraft – observers and photographs – as the most important, with flash and sound-ranging third and fourth, in part because he believed them of "limited usefulness during mobile operations."[38] He was no slave to technology; he experimented with using runners and carrier pigeons. His focus was on the effects of the system.[39] McNaughton's own predilection for new equipment and innovation is well-known, and his willingness to apply science in pursuit of greater effectiveness has earned him recognition as one of the first to use "operational research" to analyze operations. Anachronistic as such a designation is, it speaks to the importance of the individual in shaping a climate of innovation rooted in experimentation.[40]

McNaughton's leadership might have been of less use if he was in a British corps. The stable organization of the Canadian Corps helped him overcome many of the obstacles that limited efficient employment of artillery throughout the British army. One was the rigidity of the staff system that "consigned the senior artillery officer there to an adviser rather than an executive and permitted a great deal of variation in the artillery methods that were used at each level." In the Canadian Corps, a forceful personality and an innovative thinker like McNaughton could achieve great things, particularly when his immediate superiors allowed him full reign and to effectively act on their behalf. Unlike a British formation where an army Major-General, Royal Artillery (MG, RA) was prevented from direct communications with the counter-battery groups and his opinions could be ignored by the army commander, in the Canadian Corps McNaughton exercised executive command in practice if not on paper. Related was the turnover of units and formations under corps command; familiarity with both the front and personnel from other services helped to promote effective working relationships with the CBO and its staff.

Personalities played a key role in this regard. From December 1916, the GOC, RA of the Canadian Corps was Major-General "Dinky" Morrison; he recommended McNaughton for the post of CBSO. He gave McNaughton his head, a situation recognized by others. Crerar credited McNaughton as the spirit behind the development of scientific gunnery, noting during an interview in the 1960s that "the development – to the highest degree – of Counter-Battery work was entirely due to McNaughton. 'Dinky' [General Edward] Morrison, though admirable in other respects, was 'Boer War' in matters of gunnery techniques."[41] McNaughton called him a "dear old gentleman" who had become an administrator.

The other potentially contentious relationship was with the GOC, Heavy Artillery (HA), Brigadier-General R.H. Massie. McNaughton credits Massie's illness and temperament for making a potentially unworkable relationship work, and indeed Massie barely registers in many studies of the heavy artillery, not even rating the use of the initials of his christian names in Swettenham's biography of McNaughton. On several occasions, McNaughton observed that there were many points of potential division: a field gunner commanding the heavies; the awkward command structure; a relatively junior officer elevated to a role usually associated with a higher rank; and even the national divide between the British and the Canadians.[42] And his actions upon being appointed GOC, RA in late 1918 suggests that he believed that his autonomy and success as CBO was, in part, a fortunate accident: he refused to countenance the same arrangement he had worked with and instead had the CBO subordinated directly to the GOC, HA. He indicated this was a result of the demands of mobile warfare, but it also highlighted his understanding of the effort and peculiar confluence of personalities and events required to make the arrangement work. With Massie and Morrison's acquiescence, McNaughton could, when necessary, act as if the three positions were one and the same.

Leadership, circumstances, organizational climate and personalities – all of these factors mattered, particularly during the formative period in 1917. The formalization into a CBO of what had been practices in many corps caused some problems, illustrating, in particular, the importance of personality and corps' culture. The arrival of the CBSO in VI Corps upset a delicate balance created by the benign neglect of the GOC, RA, the unfortunately named Brigadier Johnny Rotten, who had allowed his intelligence officer, a Captain Harold Hemming famous as an innovator with the flash-spotting system, to experiment and communicate direct with the corps' batteries. The new CBSO promptly dismissed the work of his intelligence officer, observing that the batteries would only fire at what they could see; information gathered by other means would be binned and Hemming could "go away and stay away."[43] Not typical, but an indication that the utility of the CBSO was not consistently recognized across the British armies. In contrast, as noted above, McNaughton immediately recognized the value of what his fellow McGill alumni, and Canadian, Hemming was doing. Their mutual interest began a life-long friendship, and one that proved invaluable to the development of the Canadian counter-battery's effectiveness.[44]

None of the circumstances that afforded McNaughton the opportunity for improving counter-battery techniques would have mattered if he had not achieved something. And at Vimy Ridge, he achieved a spectacular success; the counter-battery fire accounted for 83 per cent of the 212 German guns identified, a statistic that was given some prominence then and since as critical to the Canadian's victory. McNaughton built on this success. The establishment of the CBSO had standardized the main task of the heavy artillery and, to a degree, organization, across the British army. Each CBSO had a minimal staff, with an orderly officer and clerks, and later an aide-de-camp. McNaughon's ambitious use of the developments in counter-battery techniques and practices in the planning for the operation against Vimy Ridge led to an expansion of that counter-battery office in the Canadian Corps to thirteen by the time the corps attacked, a number was that later deemed insufficient. The expansion of counter-battery work and the growing numbers of sensors, and consequent demands on communication, also led to the creation of the Corps Heavy Artillery Signal Section. McNaughton also proceeded to establish the CBO as the primary repository of corps' intelligence, and ensured it was sent up as far as the army RA and Intelligence and down the change of command to the batteries. Under McNaughton, the CBO also became an agent of inter-service co-operation, with dedicated air force resources and with direct links that ran both ways, to the RAF airplane and balloon units.[45] He was entrusted with dedicated Counter-Battery Groups. And early successes like that at Vimy Ridge only confirmed that the trust was not misplaced, while most formations in the British army struggled with doubts and divisions through to the end of 1917. For the capture of Hill 70, he had 111 guns in three groups, and access to another Heavy Artillery Group.[46]

After Vimy, his work served a model for some, and as motivation for others. Few British corps seemed able to match the advantages offered a CBO by the unique circumstances in the Canadian Corps, but to say so definitively would require a fuller study.[47] There is anecdotal evidence, some noted above, that suspicion lingered with regards to the staff position and the science of counter-battery work. The turnovers within corps did not help, although some British corps took to leaving the CBO staff behind when the headquarters moved to mitigate the impact of the constant turnover within corps.[48] There were also set-backs during the Battle of Third Ypres, not least as conditions on the ground and in the air demonstrated the limits of the intelligence-based CBO, but successes in the Battle of Cambrai on November seem to have blunted the last resistance; from that point the British CBO's began to register considerable successes.[49] The newly arrived Americans would see the Canadian Corps' and British III Corps' CBOs as models.

Within the Canadian Corps, McNaughton soon embodied, as well as benefited from, the spirit of innovation that characterized the corps headquarters.[50] His leadership provided an example throughout the corps. If the litmus test of the willingness to experiment was the commitment of resources, then the examples mentioned above suggest his success at the corps headquarters. Elsewhere, we can see examples of the techniques and organization used by the CBO spread throughout the corps. In August 1917, Harry Crerar spent most of a 10-day attachment as a "learner" with the 4th Canadian Divisional Artillery headquarters working with McNaughton to organize the six-inch Newton batteries of the Trench Mortar Brigade into counter-mortar units for use in the attack on Lens.[51] In 1918, Crerar examined and applied the counter-battery mortar organization to the 5th Canadian division's artillery. In his own recollection, he improvised a unit "through a bit of organization and by 'scrounging' and borrowing signals equipment and personnel to establish the necessary communications." Precedents set at the headquarters were important; he was able to draw on precious resources to experiment with a weapon often treated as the "Cinderella" of the artillery and the infantry.[52] And here was the beginning of one of the more important relationships in the history of the Canadian army.

The peak of CBO organization and effectiveness was reached during the fall of 1918 when the advances of the Hundred Days and the demands of mobile warfare highlighted the limits of the sensor and communication network established through 1917 and 1918. Information was supplied by air (observation aircraft and balloons), through photographs, flash spotting, range sounding, observation points, from the front lines, and prisoners of war; a 12 line switchboard had direct lines to its sources of information, and was supplemented by wireless and dispatch riders.[53] The mobile warfare of the autumn diminished the effectives of many sources of information, but practices and processes built up by the Intelligence Branch of the CBO made it, almost by default, the "Corps Advanced Report Centre."[54] The expansion of the staff and its influence reflected the growing importance, indeed the centrality, of the intelligence function of the CBO. It became the hub of a network of sensors, and it systematically collected the information and then provided intelligence, not just to the artillery, but to the corps in general. The success of the CBO was rooted in an information revolution, of sorts. And this was rooted in the static conditions of the western front. The success of this information and intelligence war also shaped counter-battery tactics. Destruction gave way to neutralization, and the role of the CBO expanded as it developed new tactics to support the infantry once the assault was underway.

The accumulated understanding of counter-battery organization, tactics and procedure was captured in two reports, one for "trench" warfare and one for "mobile" warfare, produced at the end of the war, by the second Canadian Corps' CBSO, Harry Crerar. They illustrated how much the CBO had become the primary organization for the collection, collation and distribution of intelligence for the Canadian Corps, a feature that, while not distinct, seems to have been taken advantage of by Currie and his commanders in a way not evident in other corps headquarters. Crerar suggested that the "outstanding feature" of the functions of the counter-battery office during the operations of 1918 was its "increasing importance as an intelligence centre."[55] This development owed something to the culture, but also to the degree to which McNaughton was able to demonstrate the utility and flexibility of the CBO and its main staff officer; when conditions called for the decentralization of the artillery resources, the reverse of the trajectory of the preceding two years, the CBO was still perceived as an essential component of the final operations. And it was this adaptability that was perhaps the final testament to his leadership.[56]

The period between September 1916 and April 1917 was probably the most important of the war in terms of setting the tenor of the development of the Canadian Corps; the same may said for all British formations on the western front. It was a turning point in the understanding of how to best organize and synchronize the innovations in tactics, equipment and technology that had emerged over the preceding two years; the armies began to think in terms of weapons systems. This was reflected in the changes to corps headquarters that occurred during the winter of 1916-1917 when the British Army created the Counter-Battery Staff Office, and laid the foundation for a war-winning artillery-based approach to operations. As a corps-level agency responsible for the integration of artillery intelligence with operational planning, the CBO was the manifestation of a number of inter-related developments. Still, it was in the Canadian Corps in general, and the CBO in particular, that the full potential of those developments was realized.

Andrew McNaughton should receive much of the credit for this result. The climate established by 1917 at Canadian Corps headquarters was a critical element in McNaughton's success. It was one enabler of the development of an effective artillery command system; the CBSO was treated as a commander in the Canadian Corps. This was in part a result of the personalities involved, but also a reflection of how the culture of the corps rewarded utility and effectiveness. Innovators were rewarded with promotion, and allowed a degree of autonomy not evident in comparable formations. McNaughton is a good example of how the system worked, and the factors that allowed it to work so well.[57] But he was also a leading practitioner; his early successes, particularly at Vimy Ridge, were essential to cementing this climate within the corps. Leadership and effects mattered because they were allowed to matter more than command and staff protocols, a hallmark of any organization dedicated to learning and innovation. And success built on success. In the BEF, with its five armies and 19 corps, the problem was the lack of enforceable uniformity and standardization, particularly as divisions circulated among different corps. From the perspective of the development of artillery tactics, the ambiguous command and control structure of the artillery exacerbated the problem.[58] The same autonomy and discretionary authority that corps headquarters had to distill and disseminate lessons undermined the efficiency of the BEF as a whole, but served the CEF well, tied as it was to corps' climate. The Canadian Corps was imbued, from the top down, with a willingness to learn, to innovate and to reward merit. The stability of the headquarters staffs and command from the fall of 1916 only added to the advantages.

As a reflection of leadership, the story of the CBO illustrates the maxim that leadership is situational. The establishment of the CBO may have been the result of a British initiative, but the evolution of the position was dependent on a number of factors, not least of which was the degrees to which the CBSO grasped the opportunities afforded by concurrent developments and was able to negotiate an ambiguous position. Still, the traits that resulted in McNaughton's success in the Canadian Corps – a wide-ranging curiosity, a forceful and focused personality, an emphasis on the most effective practices irrespective of rank or precedent – were among those that critics cited as problematic when he commanded First Canadian Army during the Second World War. Caution is necessary when comparing two dissimilar situations – and is interesting to speculate as to whether the traits that worked against McNaughton in the politically charged atmosphere engendered by years of training and waiting for deployment in the United Kingdom would have served him better as an army commander in operations – but it also makes the point that context is critical when evaluating the relationship between leadership and innovation.

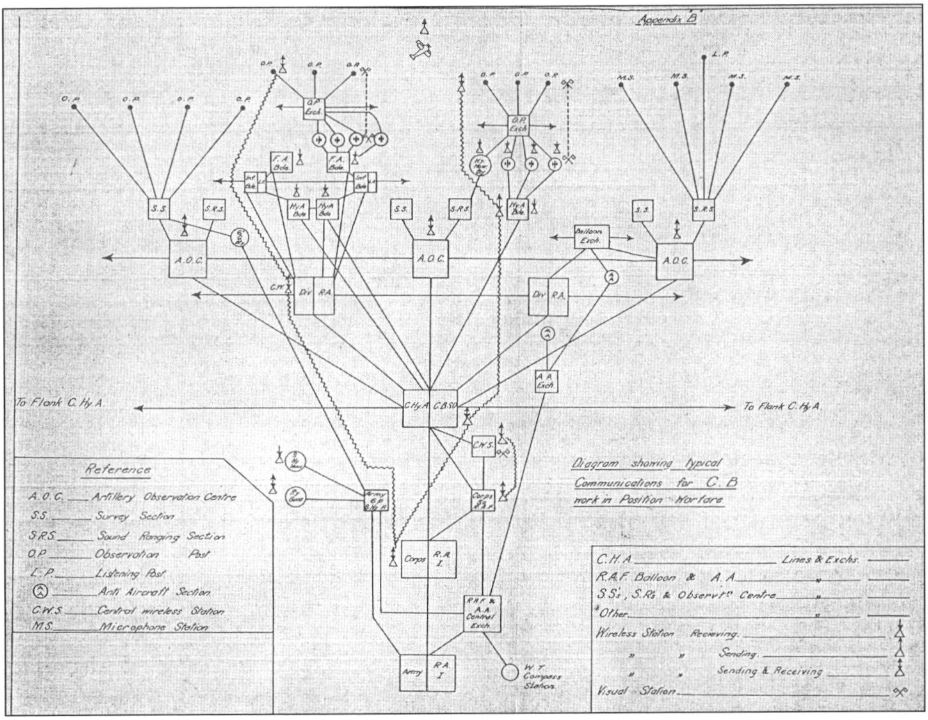

ORGANIZATION CHART, CANADIAN CORPS ARTILLERY. FROM: *REPORT OF THE MINISTRY OVERSEAS MILITARY FORCES OF CANADA* (LONDON: PRINTED BY AUTHORITY OF THE MINISTER, OVERSEAS MILITARY FORCES OF CANADA, 1918).

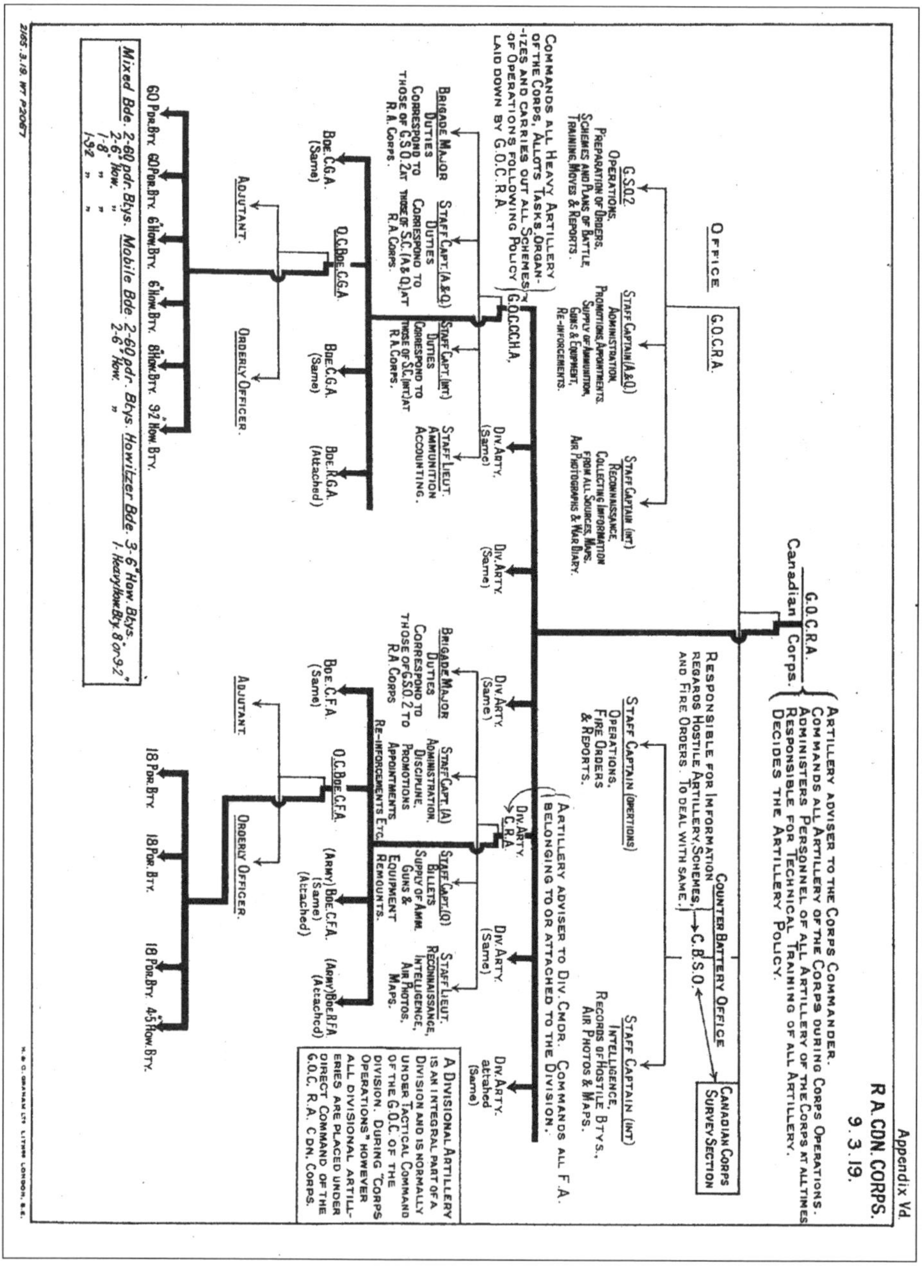

FIGURE: FROM A.G.L. MCNAUGHTON, "COUNTER-BATTERY WORK," *CANADIAN DEFENCE QUARTERLY* (1926).

1 Patrick Brennan and Thomas Leppard, "How the Lessons were Learned: Senior Commanders and the Moulding of the Canadian Corps after the Somme," *Canadian Military History Since the 17th Century: Proceedings of the Canadian Military History Conference*, May 2000 (Ottawa: National Defence, 2001) p.141.

2 The latter quote from Desmond Morton, "Changing Operational Doctrine in the Canadian Corps, 1916-17," *Canadian Army Journal*, 2 (Winter 1999): 35.

3 Albert P. Palazzo, "The British Army's Counter-Battery Staff Office and Control of the Enemy in World War I," *The Journal of Military History* 63 (January 1999): 74; Tim Travers, *How the War Was Won: Command and Technology in the British Army on the Western Front* (London: Routledge, 1992) amd Travers, *The Killing Ground: The British Army, The Western Front and The Emergence of Modern War 1900-1918* (London: Allen & Unwin, 1987); Shelford Bidwell and Dominick Graham, *Fire-Power: British Army Weapons and Theories of War 1904 - 1945* (London, 1982).

4 Morton, "Changing Operational Doctrine in the Canadian Corps, 1916-17," p.38.

5 Bruce Gudmundsson, *On Artillery* (Westport, Connecticut: Praeger, 1993), pp.1-5.

6 J.B.A. Bailey, *Field Artillery and Firepower: Combined Arms Library Volume 1* (London: Taylor and Francis, 1989), p.51.

7 G.W.L. Nicholson, *The Gunners of Canada: The History of the Royal Regiment of Canadian Artillery, Volume I* (Toronto: McClelland and Stewart, 1967), p.311.

8 Edmund Blunden reflecting on the end of the first day of the Somme, as quoted in Paul Fuessell, *The Great War and Modern Memory* (New York: Oxford University Press, 1975), p.13. "...by the end of the day both sides had seen, in a sad scrawl of broken earth and murdered men, the answer to the question. No road. No thoroughfare. Neither race had won, nor could win, the War. The War had won, and would go on winning."

9 Gary Sheffield, "How even was the Learning Curve? Reflections on the British and Dominion Armies on the Western Front, 1916-1918," *Canadian Military History Since the 17th Century: Proceedings of the Canadian Military History Conference, May 2000* (Ottawa: National Defence, 2001), pp.134-1343; Andy Simpson, *Directing Operations: British Corps Command on the Western Front, 1914-1918* (Gloucestershire, UK: Spellmount Publisher, 2006), pp.63-65; Shelford Bidwell and Dominick Graham, *Fire-Power: British Army Weapons and Theories of War 1904 - 1945* (London, 1982), pp.106-107.

10 A point identified by Brigadier-General A.G.L. McNaughton, "Counter-Battery Work," *Canadian Defence Quarterly* 3 (July 1926): 382-383 and confirmed by later studies.

11 *Report of the Ministry Overseas Military Forces of Canada* (London: Printed by Authority of the Minister, Overseas Military Forces of Canada, 1918), pp.212-213.

12 Sanders Marble *"The Infantry Cannot do with a gun less": The Place of the artillery in the BEF, 1914-1918* (New York: Columbia University Press, 2003); see chapter five; Desmond Morton, *When Your Number's Up: The Canadian Soldier in the First World War* (Toronto: Random House of Canada), pp.174-175.

13 Bidwell and Graham, pp.101-102.

14 William Rawling. "Communications in the Canadian Corps, 1915-1918: Technological Prowess Revisited," *Canadian Military History*, 3 (Autumn 1994): 10.

15 Bill Rawling, *Trench Warfare: Technology and the Canadian Corps, 1915-1918* (Toronto: University of Toronto Press, 1992), pp.85-86.

16 Rawling, *Trench Warfare*, p.135.

17 McNaughton, "Counter-Battery Work," p. 380; NAC, RG9 III, Vol 3922, File 10, Folder 7, "Organization and Procedure of Counter-Battery Office," Jan 25, 1919.

18 Paddy Griffith, *Battle Tactics of the Western Front: the British Armies Art of Attack* (New Haven: Yale University Press, 1994), pp.151-152.

19 Simpson, *Directing Operations: British Corps Command on the Western Front, 1914-1918*, pp.25-30.

20 Albert Palazzo, *Seeking Victory on the Western Front: The British Army and Chemical Warfare in World War I* (Nebraska: University of Nebraska Press, 2000), pp.111-113.

21 Simpson, pp.62-64.

22 Sanders Marble *"The Infantry Cannot do with a gun less": The Place of the artillery in the BEF, 1914-1918* (New York: Columbia University Press, 2003), pp.100-121.

23 John Swettenham, *McNaughton: Volume 1, 1887-1939* (Toronto: Ryerson Press, 1968), pp.70-71.

24 Albert P. Palazzo, "The British Army's Counter-Battery Staff Office and Control of the Enemy in World War I," *The Journal of Military History* 63 (January 1999): 55-74.

25 Swettenham, p.149.

26 Captain George Monagon and Captain James Bruce Field, "The Artillery Information Service," *The Field Artillery Journal* (Sept-Oct 1919): 438.

27 Allan D. English, *Understanding Military Culture: A Canadian Perspective* (Montreal: McGill-Queens University Press, 2004), pp.10-41; Paul Johnson, "Doctrine is Not Enough: The Effect of Doctrine on the Behavior of Armies," *Parameters* (Autumn 2000): 30-39.

28 Marc Osborne Humphries, "The Myth of the Learning Curve: Tactics and Training in the 12th Canadian Infantry Brigade, 1916-18," *Canadian Military History*, 14 (Autumn 2005): 15-30; Paul Dickson, "The End of the Beginning: The Canadian Corps in 1917," in Geoffery Hayes, Andrew Iarocci and Mike Bechthold, *Vimy Ridge: A Canadian Reassessment*, (Waterloo: Wilfrid Laurier University Press), pp.31-49,

29 Nicholson, *The Gunners of Canada*, pp.178-179.

30 Brennan and Leppard, "How the Lessons were Learned: Senior Commander and the Moulding of the Canadian Corps after the Somme," p.141.

31 Hyatt, *Currie*, pp.51-52.

32 CP, Vol 9, D219, The Development of Closer Relations Between the Military Forces of the Empire, 30 Jan 1926.

33 C.P. Stacey, "The Staff Officer: A Footnote to Canadian History," *Canadian Defence Quarterly*, 3 (1973/74): 46-47; Stephen Harris, *Canadian Brass: The Making of a Professional Army, 1860-1939* (Toronto: University of Toronto Press, 1988), pp.127-130.

34 CP, Vol 20, Crerar to Lt-General Sir Otto Lund, 17 Oct 1952; King's College of London, Liddell Hart Centre (LHC), Alanbrooke Papers, 3/A/1 Vol 1, 60.

35 Kenneth Eyre, *Staff and Command in the Canadian Corps: The Canadian Militia 1896-1914 as a Source of Senior Officers*, Master of Arts Thesis, Duke University, 1967, pp.112-115.

36 NAC, Currie Papers, MG30, E100, General Sir Arthur Currie, "Notes on French Attacks North-East of Verdun in October and December 1916" January 1917; A.M.J. Hyatt, *General Sir Arthur Currie: A Military Biography* (Toronto: University of Toronto Press and the Canadian War Museum, 1987), p.65.

37 Swettenham, Vol I, pp.70-71.

38 Brigadier-General A.G.L. McNaughton, "Counter-Battery Work," *Canadian Defence Quarterly* 3 (July 1926): 382-383.

39 Nicholson, *The Gunners of Canada*, p.315.

40 J. S. Finan, W. J. Hurley, "McNaughton and Canadian Operational Research at Vimy," *The Journal of the Operational Research Society* 48, 1 (January 1997): 10-14.

41 NAC, Crerar Papers (CO), Vol 19, Crerar's response to "Questions For a Programme on Artillery Tactics," Jan 1962.

42 Swettenham, pp.136-137.

43 Griffith, pp.152-153; Swettenham, p.71.

44 Paddy Griffith, "Tactical Reform in the British Army," in Paddy Griffith (Editor), *British Fighting Methods in the Great War* (London: Frank Cass, 1996), pp.13-14.

45 National Archives of Canada, (NAC), RG9 III, Vol 3922, Folder 7, File 9, "Notes on the Work of a Counter-Battery Office", 9 Dec 1918.

46 First quote from Shelford Bidwell and Dominick Graham, *Fire-Power: British Army Weapons and Theories of War 1904 - 1945* (London, 1982), pp.100-02; Shane B. Schreiber, *Shock Army of the British Empire: The Canadian Corps in the Last 100 Days of the Great War* (Westport, Connecticut: Praeger, 2001), pp.23 and 44; Nicholson, *The Gunners of Canada*, pp.294-296.

47 Sanders Marble *"The Infantry Cannot do with a gun less": The Place of the artillery in the BEF, 1914-1918* (New York: Columbia University Press, 2003), see chapter 5; Bidwell and Graham, Firepower, pp.100-103.

48 Marble *"The Infantry Cannot do with a gun less"*, pp.95-121.

49 Albert P. Palazzo, "The British Army's Counter-Battery Staff Office and Control of the Enemy in World War I," *The Journal of Military History* 63 (January 1999): 68-70; Simpson, *Directing Operations: British Corps Command on the Western Front, 1914-1918*, pp.117-123.

50 NAC, CP,Vol. 22, Memorandum on Artillery Notes on Operations of the Canadian Corps, 26 Aug-4 Sept 1918, 6 Oct 1918.

51 Swettenham, *McNaughton, Volume I*, 101; NAC, A.G.L. McNaughton Papers (MP), MG30 E133,Vol 108, Duguid to Crerar, 31 Aug 1922.

52 NAC, Crerar Papers (CP),Vol 19, Crerar to Grattan O'Leary, 1 Aug 1940; RG9 III C4,Vol 4283, File 9, G 50/17/660 5th Can. Div. Arty., 26 April 1918.

53 NAC, RG9 III, Vol 3922, Folder 7, File 9, "Notes on the Work of a Counter-Battery Office", 9 Dec 1918.

54 NAC, RG9 III,Vol 3922, Folder 7, File 9, CB831, CSBO to GOC, CCHA, 9 Dec 1918; Nicholson, *The Gunners of Canada*, p.317.

55 NAC, RG9 III,Vol. 3922, Folder 9, File 10, Lt. Col. HDG Crerar, CBSO to GOC, RA, 25 Jan 1919.

56 NAC, CP,Vol. 22, Memorandum on Artillery Notes on Operations of the Canadian Corps, 26 Aug-4 Sept 1918, 6 Oct 1918.

57 NAC, RG9 III, Vol 3922, File 10, Folder 7, "Organization and Procedure of Counter-Battery Office, Jan 25, 1919."

58 Denis Winter, *Haig's Command: A Reassessment* (London:Viking Press, 1991), pp.146-147.

CHAPTER 7

Lieutenant-Colonels Glen Campbell and Andrew T. Thompson and the Evolution of Native Canadian Participation during the First World War

TIMOTHY C. WINEGARD

The outbreak of the First World War in the summer of 1914 shattered almost 100 years of relative peace in Europe. Its great nations had circumvented any large-scale conflict since the defeat of Napoleon in 1815 through treaties, alliances and an aspiration to maintain a balance of power in Europe and empire. In place of war, European armies were deployed to the fringes of empire to gain territorial acquisitions within the imperial scramble, or to quell indigenous rebellions in existing colonies. By 1914, the pan-European Empire, led by Britain and France, covered 84 per cent of the globe as compared to 35 per cent in 1800.[1] Britain alone occupied one-fourth of the earth's landmass. The subjects and wards of King George V were as diverse as their environments and included Zulus, Xhosa, Maori and the Aboriginal populations of Australia and Canada.[2] Within the social norms of the Victorian era and the ethnocentric ideologies of Social-Darwinism, indigenous peoples were seen as an unfortunate component of the "white man's burden."

At the outbreak of war no European imperialist state, save for France, regarded their colonial indigenous populations as a source of military manpower.[3] Contemporary science, social biases and public opinion accepted that certain identifiable ethnic groups lacked the intelligence, discipline and integrity to fight modern war. It was believed that, since these groups were also the subjects of the vast overseas empires of the European belligerents, prudence warned against allowing them to fight and kill a white adversary in a European war, thus forfeiting white racial supremacy. War was the business and privilege of civilized races. Ideologically then, another European conflict in a chronology of many, was initially intended to be a white man's war. However, as the war accelerated and attrition became the order of the day, this gentleman's agreement soon became moot.

In reality, French colonial soldiers from Morocco and Algeria entered the front lines of France in August and October 1914, respectively.[4] Units of the British Indian Army were on the Western Front as early as October 1914 and were transferred to Mesopotamia through to Gallipoli in 1915. The First Maori Contingent from New Zealand had been active in Egypt since late March 1915 and landed at ANZAC Cove on the Gallipoli Peninsula on 3 July 1915.[5] Britain, Germany and South Africa all "recruited" Native labour, and to a lesser extent combat troops or *Askari*, during the gruelling 1914-1915 campaigns in Africa. With the mounting casualty figures and the increasing demand for auxiliary units, such as transport and service and support, both sides began to utilize their colonies as a source of men and materials.

On 8 October 1915, all Governors-General and Administrators of British Dominions and Colonies received a memorandum from the Colonial Office: "The Cabinet have asked for a report as to the possibilities of raising native troops in large numbers in our Colonies + Protectorates for Imperial service. What is wanted is an estimate of the numbers that could be raised; the length of time needed for training; an opinion as to their fighting value; and any pertinent remarks on such points as climactic restrictions on their employment, the influence of religion…+ the difficulty of officering."[6] The war was becoming a global conflict with men directly engaged from the far reaches of empire, including Native Canadians.

Approximately, 4,000 Native Canadians served in the Canadian Expeditionary Force (CEF) during the First World War, out of a total population of 103,774, representing 35 per cent of the Native male population of military age.[7] While no unit in the CEF consisted solely of Natives, both Lieutenant-Colonel Glen Campbell's 107th "Timber Wolf" Battalion (Winnipeg, Manitoba) and Lieutenant-Colonel Andrew T. Thompson's 114th "Brock's Rangers" Battalion (Haldimand

County/Six Nations Reserve, Ontario) were roughly 50 per cent Native in constitution – the highest in the CEF. The historically neglected chronicles of these two unique commanders, and their battalions, are invariably coupled to the participation of Indigenous Canadians during the First World War. Once the foundation of this involvement is established, the exploits of both Campbell and Thompson can be placed in their legitimate context.

BACKGROUND

At the outbreak of the First World War, Native Canadians did not have the rights and responsibilities of citizenship. Therefore, the Government of Canada could not, and did not, expect them to take up arms in a foreign war. Native Canadians remained wards of the Crown, and the British government rightly agreed stating that, "such an appeal to all the scattered remnants of tribes throughout the immense domain, and in varying degrees of civilization, would be practically impossible."[8] This attitude was also shared by the Canadian Ministry of Militia: "While British troops would be proud to be associated with their fellow subjects, yet Germans might refuse to extend to them the privileges of civilized warfare."[9]

Furthermore, during the negotiations of Treaties 1-6 (1871-1886) – which covered roughly the southern half of the provinces of Manitoba, Saskatchewan, Alberta and western Ontario – Native Chiefs specifically asked about military service. In October 1873, during the discussions of Treaty 3, governmental representative, Alexander Morris, was asked by an Ojibwa Chief from Fort Frances, Ontario that, "If you should get into trouble with the nations, I do not wish to walk out and expose my young men to aid you in any of your wars." To this Morris replied: "The English never call Indians out of their country to fight their battles."[10] Morris echoed this sentiment to Cree Chiefs at Fort Carlton and Fort Pitt, Saskatchewan in August 1876 during consultations for Treaty 6A: "I assured them, you will never be asked to fight against your will; and I trust the time will never come of war between the Queen and the great country near us [U.S.A.].... My words, where they are accepted are written down, and they last; as I have said to others, as long as the sun shines and river runs."[11]

As a collective, treaties were signed, not by Canada, but in the name of Queen Victoria; thus, Native nations saw treaties as an allegiance with the Crown through

Canada, but not with Canada itself. Such is exemplified by a letter to the Superintendent General of Indian Affairs, politician and poet Duncan Campbell Scott, written in August 1914 by Chief F.M. Jacobs of the Sarnia, Ontario, Chippewa Reserve, stating that his people were willing to offer "help towards the Mother Country in its present struggle in Europe. The Indian Race as a rule are loyal to England; this loyalty was created by the noblest Queen that ever lived, Queen Victoria."[12] This belief was not only guided by treaties, but also by King George III's Royal Proclamation of 7 October 1763. The Royal Proclamation outlined the relationship between Natives, proprietorship of land and the responsibilities of the Crown. It stated that Indian Affairs and any political activities between Native nations and the state remained the responsibility of the Crown; thus, it was a historic recognition of Native nations' status as sovereign, independent nations.[13]

In August 1914, Native men rushed to recruiting depots for reasons other than loyalty to the British Crown. Although the warrior ethic had stagnated as a result of residential schooling, religious education and isolation on reserves, it had not been completely repressed. While many joined, as did their white comrades, for money, adventure and employment, scores of other Natives enlisted to revive the warrior tradition and gain social status within their communities. They had heard the stories of their forefathers' exploits, while they themselves were confined to reserves. War in Europe seemed a feasible means to circumvent government policies and the Indian Act, while offering a freedom and escape from docile reserve life. Mike Mountain Horse, a Blood from Alberta, joined the 191st Battalion in 1916 with his brother Joe after their older sibling Lieutenant Albert Mountain Horse died on route to Canada having been gassed during the 2nd Battle of Ypres (April 1915). Mike served with distinction throughout the war, reaching the rank of Sergeant and receiving the Distinguished Conduct Medal (DCM) for bravery in the field:

> From the outset of this colossal struggle the Red Man demonstrated his loyalty to the British Crown in a very convincing manner.... My uncle, Chief Bull Shield, had been a great warrior of the plains. But the war proved the fighting spirit of my tribe was not quelched through reservation life. When duty called, we were there and when we were

> called forth to fight for the cause of civilization, our people showed all the bravery of our warriors of old.[14]

Indian Reserves, which were born into Crown protocol during the 1830s, served two strategic purposes: they opened up land for unhindered settlement and industry, and they also established a framework by which the Native population could be integrated or, if not, at least easily monitored. Embodied in this agenda was a construct of military training through residential schools and local militia units, although it was by no means uniform throughout Canada. As early as 1896, a request was submitted by William Hamilton Merritt, an honorary Chief of the Six Nations Reserve, to the Ministry of Indian Affairs proposing the formation of a regular unit composed of and reinforced by Native boys from Canada's residential schools in:

> ...the form of a permanent Imperial Corps recruited from our Indians... I would consider it very kind if you could ascertain from the Principals of the Industrial Schools how they would view the project of a certain picked proportion of their boys being drafted into a regiment on completing their education at the school, and how many it is likely could be supplied each year to recruit such a regiment if established. It has been held that the natural inborn instincts of the Indian lad suits him admirably for the profession of soldiering...I suppose you would consider that there would be no difficulty in recruiting a regiment from the Indians of Canada who could speak English, and who are qualified to make excellent soldiers, and who have proved themselves to be true and Loyal Subjects of the British Crown.

Although Merritt went so far as to lobby the British War Office in London, the proposal was dismissed for two reasons. The first was the financial resources required to educate the boys only to send them off to soldier around the world. Exporting educated, assimilated and "worthwhile young Indians" would inherently deplete that same cast within Canada; thus, diminishing the impact they could have on other members of their communities. Secondly, it was deduced that the scheme would appear as a draft, without Native approval, and could violate treaty agreements.[15] Nevertheless, certain residential schools did institute cadet training, such as Elkhorn Industrial School in Manitoba and St. Paul's School for Indian Boys in Alberta, where the three Mountain Horse brothers, as well as other young Native males received formal military drill and instruction.[16] During the First World War, Glen Campbell would again approach the concept of using residential schools as a source for Native Canadian recruits with greater success.

In addition to cadet training through the medium of the residential school system, there were also attempts to form complete Native militia units in the decades prior to the First World War. Although the United States had formed all-Native units as early as the Civil War (1861-1865) through to General John Pershing's Punitive Expedition to capture Mexican revolutionary Pancho Villa in 1916, Canada had no similar background.[17] The earliest attempts at forming an all-Native militia unit in Canada coincided with the 2nd South African War or Boer War (1899-1902).[18]

In September 1866, the 37th Haldimand Battalion of Rifles was formed within the active Canadian Militia with headquarters at Dunnville, Ontario (neighbouring the Six Nations Reserve) by regimentation of independent companies. It was composed of six companies of which four (Dunnville, Caledonia, Oneida and Walpole) were entirely Native, save for a majority of white officers. Given the high proportion of Natives, in February 1896 Six Nations Chief Josiah Hill petitioned the Ministry of Indian Affairs, through his regional Superintendent, Captain D.E. Cameron, to form a regiment on the reserve:

> The Six Nations Indians feel on account of their loyalty to the Crown for over one hundred years, and having fought side by side with Soldiers of the Crown in the War of Independence and the War of 1812-1814 that a Regiment bearing the name of the Royal Six Nations whose rank and file shall be composed of Indians with the head quarters at the Council House on the Grand River Reserve be established, over 1000 men can be raised if necessary.... To be commanded by Capt. Cameron our popular Indian Agent.[19]

Cameron proceeded to forward the request to his superior Duncan Campbell Scott complete with drawings of a proposed regimental uniform (kilt with Native headdress), colours and a *guidon* with battle honours of the War of Independence and the War of 1812.[20] Enthusiastic, Scott approached the Deputy Minister of

Militia, Colonel Charles E. Panet: "I may say that this Department heartily endorses the proposed movement and is prepared, should the idea be favourably entertained by your Department, to encourage and assist the Indians in carrying out the proposal to a successful issue." Although the idea was submitted to John Hamilton-Gordon, Governor-General of Canada, in March, it was rejected on the grounds of funding and interdepartmental logistical complications.[21]

With the advent of hostilities in the Transvaal in 1899 and Canada's formal commitment of an expeditionary force on 13 October 1899, Six Nations Chief Josiah Hill sent a letter directly to Queen Victoria on behalf of the sovereign entity of the Iroquois Confederacy: "I humbly beg herewith to transmit to Your Most Gracious Majesty a decision of the Chiefs of the Six Nations Council...offering Your Majesty a contingent of Chiefs and Warriors, officered by Indians or those in connection with them to serve Your Majesty in the Transvaal, in conformity with the customs and usages of their forefathers and in accordance with existing Treaties with the British Crown."[22] Also, during November, requests to send warriors were submitted through local Indian Affairs Superintendents by the individual Chippewa (Ojibwa) Reserves of Saugeen, Nawash and Sarnia (*Aamjiwnaang*), Ontario and by the Duck Lake Cree and Ojibwa of Saskatchewan.[23]

With requests flooding in from across Canada, both the British Colonial Office and the Government of Canada finally dismissed the requests. British Colonial Secretary, Joseph Chamberlain, replied directly to Chief Hill on behalf of the Queen in February 1900: "I have received Her Majesty's comment to desire you to convey to the Chiefs of the Six Nations our extension of Her sincere thanks for the loyalty and sympathetic assurances contained in the resolution and of Her regret at being unable to avail Herself of their patriotic offer." Shortly thereafter in April, the Ministry of Indian Affairs served notice to all Reserve Superintendents that, "no Treaty Indians can enlist for service," as there were reports and rumours circulating of Natives, mostly in western Canada, "wishing to join the Boer force in the Transvaal" out of sympathy for the Boers' repression at the hands of the British. There was also a fear that the provision of military training and the organization of Native regiments could in some form be utilized against Canada itself.[24]

The most palpable example of the British/Canadian policy of barring Natives from service is that of John Brant-Sero, a Mohawk from the Six Nations Reserve: "I have just returned from South Africa, disappointed in many respects, but I do not wish these lines to be understood as a grievance. I went to that country from Canada hoping that I might enlist in one of the mounted rifles; however, not being a man of European desent [sic], I was refused to do active service in Her Majesty's cause as did my forefathers in Canada.... I was too genuine a Canadian."[25]

Although the government banned the enlistment of Natives from service in South Africa, a limited number did manage to evade this protocol, and the stipulations of the Indian Act, to participate in the Canadian campaigns of the Boer War. Due to the fact that enlistment was on an individual basis and no account of "race" was registered on any formal military records, the precise number of Natives who served is not known. However, the archival and historical record does mention the exploits of a select number of Native men who were part of the 7,368 Canadians to see service during the Boer War. One such soldier was Private George McLean of the Okanagan Nation of British Columbia who served with the 2nd CMR in South Africa. McLean would again enlist during the First World War, being awarded the DCM during the Battle of Vimy Ridge (9-12 April 1917). His citation states that "single-handed he captured 19 prisoners," whilst being wounded.[26]

THE FIRST WORLD WAR

In 1914, as with the Boer War, official government policy forbade the voluntary enlistment of Natives into the Canadian Expeditionary Force despite the fact that many were active in militia units. Nevertheless, many Native men applied for overseas service. Most were immediately rejected, with many others released after their Indian status was discovered. However, under the "call to arms" of the Minister of Militia, Sir Sam Hughes, existing militia units recruited directly from their regions without interference from the Ministry of Militia or Indian Affairs. Hence, local militia officers had absolute discretion over whom they enrolled.[27] Although "race" was not recorded on enlistment documents, some recruiting officers listed "Indian" under the section entitled "Description of [Name] on Enlistment—Complexion" on the attestation form.[28] Many Natives were able to circumvent official policy, with or without the collusion of their commanding officers. The 1st Canadian Division, which disembarked in England on 14 October 1914, did contain Native soldiers, including famed Ojibwa sniper Corporal Francis Pegahmagabow, as did the independent Canadian Princess Patricia's Light Infantry Battalion. According to R.F. Haig of the Fort Garry Horse, some British civilians were disappointed to witness that the newly arrived colonial soldiers from Canada were not all red-skinned, decorated and dressed in feathers and pelts, wearing traditional headdress.[29]

Although Native Canadians were campaigning and dying in France and Belgium by early 1915, official policy still denied them enlistment privileges.[30] Shortly after the declaration of war in August, there were numerous unsuccessful attempts to create all-Native units throughout Canada. The first effort was made in October 1914 by Glen Campbell, Chief Inspector of Indian Agencies for western and northern Canada, and militia veteran of the 1885 Northwest Rebellion. Similar to the mounted infantry unit in which he confronted Métis and Cree forces during Louis Riel's second uprising, Campbell proposed the formation of a corps of scouts or irregular cavalry made up entirely of Natives. Concurrently, in Alberta, The Reverend John McDougall, a Methodist missionary to the Alberta Natives, petitioned the Ministry of Indian Affairs suggesting that: "Indians at one time fought in battles amongst themselves, and some of them are the best scouts in the world.... I would suggest the taking of a certain number of Indians from each tribe and from each reserve, and making up a regiment of about 500."[31]

Similarly, in November 1914, Colonel William Hamilton Merritt, honourary Chief of the Six Nations Iroquois, tried, as he had done during the Boer War, to arrange with the Ministry of Militia the formation of a Six Nations battalion, which he personally offered to fund. The Militia Council in Ottawa replied that his offer was simply "too inconvenient."[32] All attempts at creating all-Native units or sub-units, including Campbell's, were rejected conjointly by the Ministry of Indian Affairs and the Ministry of Militia throughout the early months of the war. In December 1914, Duncan Campbell Scott bluntly stated that, "no unit composed solely of Indians will go to the front with the Canadian Contingent."[33]

Despite this policy, by late 1914 and into early 1915, scores of requests were being sent by Band Councils, Indian Agents and private citizens to both the Ministry of

Militia and the Ministry of Indian Affairs regarding the enlistment of Natives or the formation of Native units. In response, the Ministry of Indian Affairs issued a directive in late December 1914 to all Indian Agents stating that, if men wanted to enlist, they were to be taken or shown the recruiting office, but there was to be no direct recruiting effort on behalf of the agents themselves.[34] Despite this instruction, there remained confusion among Natives, their Indian Affairs Superintendents and even District Military Commanders as to the regulations on Native enlistment. However, this ambiguity was quickly rectified by the Ministry of Militia.[35]

On 6 December 1915, the Ministry of Militia, with support from the Ministry of Indian Affairs, ceded its position against overseas service for Native Canadians. Instructions were sent to all officers commanding military units or districts to accept Natives for service so long as they met the requirements and standards of the enlistment regulations.[36] This decision was not based on an ideology of ethics or equality. It was simply a realization of the recruiting demands required to fuel the manpower needed to fight a modern war with horrific casualty rates. The 1st Canadian Division had been drained by the 1915 battles of Ypres, Festubert and Givenchy, losing 6,037 men in just four days during the 2nd Battle of Ypres alone. In addition, Canadian Prime Minister Sir Robert Borden announced in December that as of 1 January 1916, Canada would pledge 500,000 soldiers to the European fronts. In reality, assuming the prevailing casualty rates, 300,000 new recruits per annum would be required to maintain this commitment.

THE FORMATION OF THOMPSON'S 114TH AND CAMPBELL'S 107TH BATTALIONS

With the advent of authorized policy to enlist Native Canadians for overseas service, battalion commanders and recruiting officers wasted little time. The first unit officially to induct Native members was the 114th Battalion headquartered in Cayuga, Ontario, astride the Six Nations Iroquois Reserve. Since its formation on 9 November 1915, there had been continuous correspondence between its commanding officer Lieutenant-Colonel Edwy Sutherland Baxter and the Ministry of Militia concerning the enlistment of Natives, given its proximity to the reserve, and also to the fact that it was constituted from the 37th Haldimand Rifles militia unit, which contained a majority of Natives.[37] Although having initially formed the 114th Battalion, Lieutenant-Colonel Baxter's tenure of command lasted just over two months. He succumbed to disease on 15 February 1916 in Cayuga and was replaced by Lieutenant-Colonel Andrew Thorburn Thompson, born in Cayuga on 27 May 1870.

Pedigreed by an affluent, aristocratic family, Thompson had previously served in the Canadian Militia as Private to Colour Sergeant with the Queen's Own Rifles of Canada. He then joined the 37th Haldimand Rifles as a captain in 1892 and proceeded to serve as the regiment's commanding officer for eight years. He went on to command the 5th Infantry Brigade for four years and served as commander for the Canadian Coronation Contingent at the crowning of King Edward VII in January 1902.[38] Thompson, a lawyer and editor (he was editor of the Canadian Military Gazette for many years) by trade, also served as Member of Parliament for the constituency of Haldimand and Monck from November 1900 to November 1904 in Prime Minister Wilfrid Laurier's Liberal Government.[39]

Given his prior service with the 37th Haldimand Rifles and his political experience, Thompson was a logical choice to command the 114th Battalion given its Native composition. In addition, his grandfather had fought along side Six Nations warriors at the Battle of Queenston Heights with Major-General Sir Isaac Brock, and Thompson's two sons, Andrew and Walter, were serving lieutenants, having joined the 114th on 8 December 1915. Thompson himself was an Honorary Chief of the Six Nations, his Iroquois name being *Ahsaregoah* meaning "the sword."[40] In him, his soldiers found a commander with whom they were familiar and trusted. Although not participating directly in the Boer War, during Christmas of 1900 Thompson, as local MP, had urged the members of his community, the Six Nations Reserve and his 37th Haldimand Rifles to "Remember there are thousands of our war wrecked men still in hospital—VISIT THEM. Do not let them think they are forgotten, nor remember them only on Christmas Day."[41]

Commanding the 114th Battalion as of February 1916 and with the decree of 6 December 1915, Thompson was given permission to enlist Natives both within and outside the battalion's geographical area and military district. He was also afforded the opportunity to arrange for the transfer of Natives serving in other units who wished to join the 114th, with a view to forming an all-Native battalion.[42] This arrangement was lent support by Brigadier-General W.A. Logie, Commander of No. 2 District Toronto, and also Duncan Campbell Scott: "I thought I should write you [Logie] and state how much I am interested in the welfare of the 114th Battalion; I hope to see a solid half of the battalion composed of Indians, and I trust that District No. 2 may be able to produce them. It is in the interest of the Indians, I think, that we should have at least two full Indian companies. Personally and officially I have been doing everything possible to bring this about."[43]

During active recruiting and in the media, the 114th was advertising itself as "the Indian Unit," and at least a dozen regiments transferred their Native recruits (and "half-breeds" as one battalion mentioned) to the 114th. In a gesture of solidarity, the Ministry of Indian Affairs lent the "only male Indian employed at the Service in Ottawa," Charles Cooke, an Iroquois, to the 114th Battalion as a recruiting officer. Bestowed with the honorary rank of lieutenant, Cooke toured Ontario reserves throughout 1915-1916, often with Lieutenant-Colonel Thompson or "another commissioned Indian," recruiting for the 114th. By the end of his first week Cooke had recruited 90 Natives.[44] However, not all previously or newly enlisted Natives wanted to serve in the 114th Battalion. Natives from other nations specifically asked not to be transferred, as they preferred "not to fight alongside Mohawks." Historically, both pre and post European contact, the Iroquois Confederacy was the dominant Native coalition in northeastern North America. Iroquois warriors had earned a reputation as fierce combatants (specifically Mohawk) by conquering or assimilating other Native nations, some to the point of near extinction as in the case of the Huron/*Wendat* and the Mahican Confederacy.[45]

After only two months of this practice, the exercise of transferring Natives from other units to the 114th was abolished due to pressure and complaints from other battalion commanders who were also trying to fill their under-strengthened units. One battalion even offered Natives a five dollar recruitment bonus, in addition to a free trip to Europe, in case the war ended before they were shipped overseas.[46] Thus, there was a concentration of Natives in the 114th, but many others were scattered throughout the country in other battalions.[47] Nevertheless, the 114th Overseas Battalion successfully recruited 50 Mohawks from Kahnawake and

Kanesatake (Oka) in Quebec, a considerable number of Mohawks from Akwesasne/St. Regis, Quebec/Ontario/New York State, as well as a number of Natives from northern Ontario and Manitoba. In all, 353 Natives (287 from the Six Nations Reserve) served in the 114th Battalion, of which two of the four companies including most of the officers were Native, and were classified as such on battalion nominal roles. Attached to the battalion was a 35 piece regimental band, all from the Six Nations Reserve. The band toured the British Isles for recruiting and patriotic purposes and included traditional garments and war dances in their performances.[48]

Thompson wasted little time in showing his support and allegiance towards all members of his battalion including his Native soldiers. Although only two companies were comprised of Natives, special concessions were asked for by Thompson in a letter to the Ministry of Militia on 25 March 1916:

> This battalion is recruiting largely from the Six Nation Indians. Already more than two hundred of them have enlisted, and I confidently expect three hundred and fifty to four hundred more. The ancestors of these men fought for Great Britain in every battle on the Niagara frontier in the War of 1812, and were with General Brock in large numbers when he fell at Queenston Heights. To this day they venerate his memory, and the name for which I ask, 'Brock's Rangers' would greatly add to our prestige with them, and gratify them exceedingly. The 'white' half of the battalion comes form Haldimand County, one of the Niagara Peninsula group, and many of these men too had ancestors with Brock in 1812.

Permission was granted to use the name "Brock's Rangers" two days later.[49]

The regimental crest featured two crossed tomahawks below the regimental motto "For King and Country." The crest also bore the name "Brock's Rangers" and a crown, all superimposed on a maple leaf. The Six Nations Women's Patriotic League also embroidered a 114th Battalion flag, which they adorned with Iroquoian symbols. Thompson proceeded to petition and gain approval from the Ministry of Militia for his battalion to carry this flag alongside the King's Colours and their regular Regimental Colours, as the latter two banners were the only official garlands recognized by His Majesty's forces.[50] The battalion was mobilized on

29 September 1916 to proceed overseas. The last inspection before sailing was conducted on 17 October 1916 at Camp Borden by Major-General F.L. Lessard who concluded that: "This is a good Battalion in which there are 300 Indians...15 men trained in Scouting...of Good Class and Physique."[51]

The only other CEF unit to mirror the Native composition of the 114th was the 107th Battalion raised in Winnipeg, Manitoba in December 1915, shortly after governmental permission was granted to enlist Natives. Similar to the 114th Battalion, an all-Native formation was envisaged save for its officers. Lieutenant-Colonel Glen Campbell, who had attempted to construct a similar unit in 1914, was responsible for the configuration and recruitment of the battalion and became its first commanding officer. If Canadians created legendary folk heroes, like those of the United States, Campbell would be equivalent to the celebrated American frontiersman, soldier and politician Davy Crockett.

Glenlyon Archibald Campbell was born on 23 October 1863 at the Hudson's Bay Company Post, Fort Pelly, Saskatchewan. Glen's father, Robert Campbell, an immigrant from Glen Lyon, Scotland, was a fur trader for the Hudson's Bay Company (HBC) for interrupted periods between 1830 and 1871.[52] Through his trading expeditions, he helped map the final portions of northern Canada giving name to many geographical features in Yukon Territory. In fact, in 1840, he became the first white man to cross into the Yukon River watershed from the east. The present day 602km long Robert Campbell Highway (Highway 4) in the Yukon roughly conforms to his 1840s route.[53]

Robert Campbell's vocation had an early impact on his son Glen's life. In 1870, Glen and his two siblings accompanied their mother to Scotland, where shortly thereafter she died of typhoid.[54] The children were taken in by an aunt in Perthshire and spent their time between Scotland and Manitoba. In the process, Glen attended Glasgow Academy and Merchiston Castle School in Edinburgh. By 19 years of age Glen found himself in Montana Territory, at his father's request for him to work on a cattle ranch and learn stock-raising. In 1884, Glen returned to Manitoba to live with his father on the family ranch near Riding Mountain in the Russell District of Manitoba.[55]

Unlike Thompson, prior to taking command of the 107th Battalion, Campbell's only military service had come during Louis Riel's Northwest Rebellion of 1885. In early April 1885, Major Charles Arkoll Boulton was given permission from the Canadian government to recruit an irregular mounted infantry unit from the population of the Russell-Birtle District in Manitoba. The unit known variously as "Boulton's Mounted Infantry," "Boulton's Horse," and more commonly as "Boulton's Scouts," consisted of 5 officers and 123 men, including Glen Campbell. The unit joined General Frederick Middleton's column as the advance guard on route to Fish Creek and the Métis capital of Batoche. On 13 May, following the Battle of Batoche, Major Boulton promoted Campbell; the previous two troop leaders had been wounded and killed respectively: "I now appointed Captain Campbell, a son of an old Hudson's Bay officer.... He was installed amid cheers of the men."[56] In actuality, the Campbell/Riel connection dated back to the Red River Rebellion (1869-1870). In fear of Métis pillaging and reprisal, Glen's father, by then a Chief Factor in the HBC, sent his year's quota of furs to London, England via Sioux (*Lakota*) country, which was engulfed in the American-Sioux War (1862-1890). Glen, his mother and his siblings accompanied the military escort from Manitoba and disembarked in London. This action led to Robert's dismissal from the HBC the following year.[57]

After his brief period of military service, Glen Campbell returned to ranching, hunting and trapping and in 1897-98, during the Klondike Gold Rush, he unsuccessfully attempted the arduous overland route from Edmonton to Dawson City. Like his counterpart Thompson, Campbell too became a politician. After being defeated for the seat of Dauphin in the Legislative Assembly of Manitoba in 1892 and 1896, he won the reconstituted constituency of Gilbert Plains[58] in 1903 and again in 1907 as a Conservative. In 1908, aided by his brother-in-law, MP Clifford Sifton, he was elected to the House of Commons for the federal riding of Dauphin.[59] From all accounts, the 6'4" Campbell made an impression on his fellow Members of Parliament with his "towering figure, bronzed swarthy face, large brown eyes, capped off with a cowboy hat."[60] Although he spoke infrequently, on 17 March 1911, Campbell was involved in a bellicose dialogue with an Alberta MP, in a confrontation labelled "one of the tensest situations and exciting scenes ever witnessed in Canadian Parliament" by *The New York Times*.[61] On another occasion, after listening to a speech by an eastern member who used eloquent yet confounding language, Campbell made his rebuttal in a combination of Cree and Latin.[62]

Campbell, although a Conservative, was defeated in the 1911 election which saw Robert Borden replace Wilfrid Laurier as Prime Minister. However, he was appointed Chief Inspector of Agencies, Reserves and Inspectorates in Manitoba, Saskatchewan, Alberta and the Northwest Territories within the Ministry of Indian Affairs by the new Conservative administration. Campbell had strong ties with the Native communities of Manitoba through ranching and trapping. He had also married Harriet Burns, daughter of Saulteaux Chief Keeseekoowenin, in 1886. Campbell remained in this position with Indian Affairs, centred in Winnipeg, until July 1915.[63]

As mentioned, in October 1914, Campbell unsuccessfully petitioned Ottawa to raise an irregular unit of Native scouts similar to that in which he had served under Major Boulton during the 1885 rebellion. However, on 9 July 1915, at 52 years of age with virtually no military experience, he was loaned to the Ministry of Militia by the Ministry of Indian Affairs, given the rank of major and tasked to assist in raising the 79th Battalion in Brandon, Manitoba (It should be noted that he was not the commanding officer). On 24 November, Campbell was transferred from the 79th and mandated to raise the 107th Battalion in Winnipeg and was appointed its commander as a lieutenant-colonel.[64]

Having been assigned this duty and with the 6 December 1915 governmental decree to allow the recruitment and enlistment of Native Canadians, Campbell was afforded the opportunity to create a Native unit such as he had desired in 1914. However, Ministry of Militia and Indian Affairs regulations still opposed a homogenous Native formation. Despite this impediment, Campbell quickly set out to recruit as many Natives as possible into his 107th Battalion. On 3 February 1916, Campbell propositioned Duncan Campbell Scott seeking permission to recruit Natives from the Elkhorn and Brandon Industrial Schools in Manitoba. Given his affinity for, and intimacy with, Native Canadian culture, Campbell argued that these young Native men "would be under closer and more kindly supervision than in any other Battalion in the west…even if they were not quite eighteen years of age."[65] Scott endorsed the proposal with trepidation. He argued that parents and Band Councils might complain if Indian Affairs used its influence to persuade underage Native pupils to enlist, but added in paradoxical logic, that those who did so would be, "breaking their treaty obligations, as they promised to be loyal citizens and it is anything but loyal to prevent recruiting." Scott then went on to state that, "there should be some good material at Elkhorn," as the students received drilling instruction within their curriculum. He also encouraged Campbell to visit western reserves for recruiting prospects.[66]

The recruiting drive by Campbell to enlist Native soldiers into his battalion was a success. Unlike the 114th Battalion that was linked to the predominantly Native militia unit of the 37th Haldimand Rifles, the 107th did not have the benefit of such a relationship. Complete, with a pipe and drum band, the motto "Follow Me!" and the regimental march, "The Campbell's are Coming," Lieutenant-Colonel Campbell enticed 1,741 volunteers, both Native and non-Native, to join his battalion. He rejected over 600 of these men and achieved full strength within three months of the battalion's conception – 45 officers and 861 other ranks. Over 500 of these soldiers were Native. However, unlike the 114th, almost, if not all, officers were white, lending credence to the comments of the Inspector-General that the "NCOs and men are very good, of good physique and above average intelligence, though very few hold certificates."[67] For the most part, a non-battlefield commission in the CEF still required an education, which eluded most western Natives. In 1914, only 3,143 of the 10,290 Manitoba Natives (31 per cent) spoke English according to Indian Affairs, in relation to 17,744 out of 26,419 (67 per cent) in Ontario, home to the 114th Battalion.[68]

Many of the Native soldiers of the 107th did not speak, or spoke very little, English and they came from a variety of Native nations: Cree, Blackfoot, Sioux and Ojibwa from the north and west; Iroquois, Delaware and Ojibwa from Ontario and Quebec, and Micmac from the Maritime Provinces. To remedy this, Campbell often instructed training, conducted parades and administrative and disciplinary matters in Native languages, as he was fluent in dialects of Cree and Ojibwa. There is also evidence that English language instruction was given to Native soldiers in the battalion.[69] Like the Native soldiers of the 114th Battalion had in Lieutenant-Colonel Thompson, the Native warriors of the 107th, also had a commander in Campbell, who was accommodating to their needs and respected their unique cultural traits.

Akin to the 114th Battalion, the 107th also had a cap badge which embodied its Native configuration. It was composed of a crown bearing the battalion number with "Winnipeg" appearing below on a ribbon, reinforced with a backdrop of a lone, stalking wolf. The 107th unofficially became known as the "Timber Wolf Battalion." The origin of the cap badge and nickname was explained by Steven A. Bell in a footnote to his 1996 journal article, "The 107th 'Timber Wolf' Battalion at Hill 70:"

> My Grandfather was a rancher in the Canadian West during the Great War. Four of the Native Canadians who worked for him joined the 107th. Only one returned. He gave my family a 107th cap badge. He claimed the Timber Wolf was selected because it was a common totem to many of the Native soldiers. He used the name 'Timber Wolf Battalion' to refer to the unit. No other explanation regarding the origin of the cap badge was discovered in the records at the National Archives.[70]

Another theory, according to Glenlyon Campbell (the great-grandson of Lieutenant-Colonel Campbell), was that Glen's Saulteaux wife, Harriet, was of the Wolf Clan, hence; the cap badge had the representation of a timber wolf: "She was full-blooded Ojibwa, daughter of Chief Keeseekoowenin, and her family was Wolf Clan. I assume that is why the 107th had the timber wolf on their patch. Apparently, Glen also gave battlefield orders in Cree, Ojibwa, and Latin, as he was proficient in all."[71]

THE FORTUNES OF WAR

After training in Canada, which for both the 114th battalion and 107th battalion consisted of significant amounts of scouting and bayonet drills according to their war diaries, both units were given clearance to embark for England.[72] However, during this instructional period in Canada, Lieutenant-Colonel Campbell's health had taken a turn for the worse. With his son John having enlisted as a lieutenant in Winnipeg's 184th Battalion, and he himself newly married to a much younger woman, Campbell spent months in hospital suffering from a serious kidney infection. According to his great-grandson Glenlyon, "Glen had kidney problems for quite a few years before he went overseas. He had tried to stop a team of stampeding horses at a parade in Dauphin, was trampled, and never fully recovered. Allegedly, during one operation in Winnipeg, he refused anesthetic, so he could watch the surgery." Nevertheless, Campbell proceeded overseas with his unit and his son, who had recently requested a transfer from the 184th in order to serve under his father.[73]

The 107th Battalion's thirty two officers and 965 other ranks sailed from Halifax on the *Olympic* on 18 September 1916, with Campbell being the senior officer on-board, and arrived in Liverpool on 25 October. Shortly thereafter, the 30 officers and 679 other ranks of the 114th left Halifax on 31 October 1916, disembarking at Liverpool on 11 November 1916. Unfortunately for both the 114th and 107th Battalions, they would not remain intact with their Native sub-units. Like many Canadian battalions, the 114th was broken up soon after its arrival in England, its members being scattered to reinforce other units. Some Native members, primarily officers, were transferred to the 107th but the majority was sent to bolster the 35th (Toronto) and 36th (Hamilton) Battalions. In addition, the 114th's regimental band toured until the end of 1917, when it was dissolved and its members sent to various formations.[74]

Lieutenant-Colonel Thompson would never command any unit of the CEF in battle. Upon, the dissolving of his battalion, Thompson was attached to the 36th Battalion to enforce and coordinate disciplinary proceedings. From 27 December 1916 to 25 January 1917, he toured as a training advisor to various units in England. After a brief stay in hospital with urethritis, he was made commanding officer of the Central Ontario Regimental Depot in south London. Health problems continued to plague Thompson and he was admitted to hospital again on 29 October 1917 with kidney stones and severe infection. He was released from hospital in July 1918 and was returned to Canada in August. He was honourably discharged from military service on 30 September 1918 and repaired to Cayuga where he remained active in the Six Nations community and Band Council politics.

In 1923, Thompson was appointed by the federal government to investigate the political organizations and Band Council of the Six Nations Reserve after a series of confrontations and violent episodes involving rival Chiefs, their supporting factions, the Royal Canadian mounted Police (RCMP) and a traditional, sovereign Longhouse group led by Chief Deskaheh. Completed in November 1923 but not released for another nine months, Thompson's report suggested the establishment of an elected band council under the Indian Act arguing that, "The separatist party [Deskaheh and supporters], if I may so describe it, is exceptionally strong in the Council of Chiefs, in fact it is completely dominant there. Its members maintain... that not being British subjects they are not bound by Canadian law, and...the Indian Act does not apply to the Six Nations Indians." Without conference with Six Nations representatives, the government deposed the traditional hereditary Longhouse Council and, though voter participation was exceedingly small as traditional Iroquois boycotted it, a new council was elected in October 1924.[75] Four years later, Thompson, accompanied by friends, travelled throughout "the Brazils...as gentlemen adventurers," the exploits of which Thompson published privately that same year.[76] Andrew Thorburn Thompson died on 20 April 1939 at the age of 68.[77]

Although never commanding his beloved "Brock's Rangers" Battalion in combat, three of his former Native officers, Lieutenants James David Moses, Oliver Milton Martin (both Mohawks from Six Nations) and Mohawk John Randolph Stacey from Kahnawake, Quebec, all served as pilots in the Royal Air Force after stints in both the 114th and 107th Battalions. Moses was shot down and killed on 1 August 1918. Stacey, a friend of Canadian ace Billy Bishop, VC, was killed on 8 April 1918.[78] Martin went on to serve as an Infantry Brigade commander during the Second World War, attaining the rank of brigadier-general, the highest position ever attained in the Canadian Forces by a Native.[79]

The fate of the 107th Battalion and that of its commander was more auspicious than that of Thompson and the 114th. On arrival in England, the 107th was initially disbanded, with 382 men transferred to strengthen the 16th (B.C./Manitoba) and 44th (Winnipeg) Battalions and the 1st Canadian Mounted Rifles (Brandon). However, the evolution of the war necessitated the creation of supplementary specialized units, and on 1 February 1917, the 107th was officially converted to a Pioneer Battalion. Pioneers were primarily infantry soldiers trained to perform basic combat engineering assignments in the front lines. However, they were not support troops like those of Entrenching Battalions, Railway Companies or Tunnelling Companies, nor are they educated and skilled engineers like those of the Engineering Field Companies. Pioneers were armed, trained in infantry tactics and employed in combat roles when not performing minor engineering tasks.[80]

Campbell himself was adamant that his soldiers were highly suitable for such vocation and his superiors thought that the Natives would perform better under their own officers. At his request, efforts were made to transfer Native officers (most from the defunct 114th) and soldiers from other units, including all Natives who had recently been sent to bolster the 16th, 44th and 1st CMR Battalions, to Campbell's command. After augmented training in England, Campbell and his pioneers disembarked at Boulogne, France on 25 February 1917. By 1 March, they had joined the Canadian Corps in the Vimy region, as part of General Horne's British First Army. The 107th was the last complete unit from western Canada to join the Canadian Corps.[81]

Although the pioneers entered the operational theatre at Vimy on 1 March, disquieted for the safety of his men, Campbell defied orders to send work parties from his battalion to forward areas, as his soldiers had not been fitted with Box Respirators. When this vital piece of equipment was issued on 4 March, Campbell ordered mandatory gas drills for the following two days, after which time work parties from the 107th began to execute front line duties. The unit participated in the preparations for and the battle of Vimy Ridge performing such functions as burying wire and cable to ensure stable communications, raising casualty clearing stations, helping to build light railway, digging trenches and erecting defensive obstacles and barbed wire. On 9 April, the opening day of the British Arras Offensive, three companies were "under orders to assist in burying cable and improving communication forward through NO MAN'S LAND." During Canadian Corps actions at Vimy, the unit suffered 10 casualties including three killed.[82]

Having incurred another 25 casualties during the month of July, the battalion's first true combat initiation coincided with newly promoted Canadian Corps Commander, Lieutenant-General Sir Arthur Currie's first operation at Hill 70 (15-25 August 1917). Until this assault, although having held a portion of front line trench from late June to mid-July, the 107th had been exclusively used for pioneering purposes. During the attack on Hill 70, the 107th was in direct support of the 3rd Infantry Brigade, 1st Division. The first Canadian troops went over the top at 4:25 am on 15 August. The 107th followed the lead waves with a view of digging communication trenches across the 300-500 yards of No Man's Land linking the original Canadian front lines to newly captured enemy positions. A secondary task, if required, was to act as primary reinforcements to the attacking infantry units and be prepared to defend captured enemy positions against inevitable counter-attack. Under fire from German artillery, including gas shells, members of the 107th Battalion's three forward companies, carried out their prescribed responsibilities throughout 15-16 August suffering 21 killed and 140 wounded (including nine gassed) out of 600 men engaged.[83]

On the night of 17/18 August, the battalion was ordered to rotate back to the rear rest areas. One company volunteered to "search and bring in wounded," from the battlefield, an offer which Lieutenant-Colonel Campbell immediately authorized. Thirty dead were recovered and buried, while another 25 wounded were brought to dressing stations. Unluckily, German artillery laid down a gas shell barrage during the process, poisoning 84 members of the company and wounding four others. Two Native members of the 107th, Privates O. Baron and A.W. Anderson were awarded the Military Medal for bravery.[84] In addition, Campbell received a letter from Lieutenant-Colonel D.M. Osmond of the 10th Battalion on 28 August thanking the 107th for "the splendid assistance you gave us on the night of 17-18th, 1917…car[ing] for a number of our wounded."

A second letter of appreciation was sent to Campbell from Osmond's direct superior Brigadier-General Frederick O.W. Loomis, commander of the 2nd Brigade. The rescue mission was even known to Lieutenant-General Currie who wrote Campbell on 31 August: "I want to tell you that I have heard, with a great deal of pride and satisfaction, the reports of the gallantry and devotion to duty displayed by…your Battalion during the recent operations…my sincere congratulations."[85] In reality, Campbell was instrumental in maintaining his battalion's esprit de corps. Given his character, he asked nothing of his soldiers that he would not do himself and was a frequent figure in the front lines, while also regularly visiting his wounded soldiers in dressing stations and hospitals.[86]

Campbell himself believed that the elevated performance of his "Timber Wolf" battalion was due to the, "courage, discipline, and intelligence of his Indian soldiers...their ability to adapt themselves without complaint to awkward circumstances and bad weather, which rendered their efficiency as a pioneer battalion far above the average."[87] It is evident in the war diary of the 107th that its Native men were proud of their heritage and their accomplishments in uniform. In fact, curious by their absence is mention of disciplinary problems; only two charge parades were recorded in one year of entries. In order to maintain positive morale, Campbell also frequently included sporting achievements (football, running and boxing) from his and other battalions in routine orders. On 26 June 1917, he was quick to incorporate that fact that the 107th's own Private Tom Longboat from the Six Nations Reserve had won the British First Army's eight-mile run from Vimy to Arras, with Corporal Joseph Keeper, a Cree from Norway House, Manitoba placing fifth.[88]

The fortunes of the 107th and Campbell did not last. On 9 October 1917, while his unit began work for the Passchendaele offensive, Campbell's chronic kidney problems forced him into hospital. He succumbed to toxaemia caused by kidney failure on 20 October 1917, three days shy of his 54th birthday. He is buried at the Etaples Military Cemetery, France alongside 1,144 other Canadian and 9,628 Allied war dead. During his short military tenure, Campbell received two Mentions in Dispatches and was posthumously awarded the Distinguished Service Order on 1 January 1918 for his leadership in sustaining his battalion's morale. He was replaced by Major Hugh C. Walkem on 23 November 1917 as the commanding officer of the 107th Pioneer Battalion. However, Walkem's term of command was fleeting. On 28 May 1918, the 107th "Timber Wolf" Pioneer Battalion was disbanded and its members absorbed into the 1st, 2nd, and 3rd Canadian Engineer Battalions, 1st Canadian Engineer Brigade.[89]

CONCLUSION

Despite the efforts of both Campbell and Thompson, no Canadian unit formed during the Great War was exclusively Native. Similar to both the 107th and 114th Battalions, a number of other Canadian Expeditionary Force units did have a relatively high percentage of Natives, although none were comparable in numbers.

Battalions such as the 135th Middlesex (with Ojibwa, Oneida and Muncey from London, Ontario and surroundings), the 149th Lambton (with Illinois and Ojibwa from the Walpole Island, Kettle/Stoney Point and Sarnia Reserves in Ontario), the 160th Bruce (with 65 Ojibwa and Potawatomi from the Saugeen and Cape Croker Reserves near Owen Sound, Ontario), the 188th stationed at Kamsack, Saskatchewan (with Nahane from the Pelly Agency) and the 52nd "Bull Moose" Battalion (with 65 northern Ontario Ojibwa and Cree) were all dispatched overseas containing a large segment of Natives.[90]

There are varying reasons why an all-Native battalion never materialized during the First World War. The first was the availability of suitable voluntary reinforcements given the exclusion of Native Canadians from the Military Service Act (August, 1917). Given the high casualty rates and the relatively low Native male population of military age in Canada (roughly 11,500), it was considered unlikely, with valid reasoning, that enough replacements could be mustered to support such a battalion once overseas and engaged in combat. For example, the New Zealand Maori Pioneer Battalion struggled after the bloodletting at the Somme to maintain its strength. Only through volunteers, transfers and conscription did it achieve full strength (900 men) in 1917. It maintained this state for the duration of the war and was the only battalion of the New Zealand Expeditionary Force to return to the dominion as a complete unit.[91]

Secondly, there was concern within the Government of Canada, that if any all-Native unit was formed piecemeal from various Native nations, there could be, as Chief of the General Staff, Major-General W.G. Gwatkin remarked, a worry, "whether coming from different tribes, they would fight among themselves."[92] In addition, many Natives did not speak English and exercising proper command and control would be very difficult given the multiplicity of Native languages that would make up any all-Native unit. Thirdly, Native Canadians' well-documented tendency to develop tuberculosis troubled the senior staff of the Canadian Red Cross. Tuberculosis rates were 20 times higher in the Native population as compared to the rest of Canadian society.[93]

Finally, W.M. Graham, the Inspector General of Indian Agencies for southern Saskatchewan argued that "If they went into the front as a unit, and if by chance they went into action and suffered tremendous casualties, there would always be a feeling among their friends at home that their sons had been placed in a more dangerous position than the whites. Of course this would not be the case... the old Indian is quite primitive and does not understand things as we do."[94] In retrospect, Graham's insight, albeit containing elements of prejudice, was quite remarkable. Many colonial, dominion and Canadian units were seen by their home governments (who exercised little discretion over their soldiers, especially prior to 1917), as being used as "cannon fodder" by British high command during the First World War.

It must also be remembered that Minister of Militia, Sir Sam Hughes, did not endorse the concept of all French Canadian battalions either, thus alienating a great number of French Canadian men who otherwise may have enlisted. The French Canadian population comprised 30 per cent (roughly 2 million) of Canada's total population and far exceeded that of Native Canadians. Yet, their recruitment levels were the lowest of any province in Canada (five per cent of men of military age) and made up only four percent (15,000 in total of which half were English speaking Quebecers) of all Canadian volunteers.[95] Had all-Native units been formed, French Canadians would have objected or demanded the same conditions.

Although no complete Native battalion was conceived during the First World War, the 107th "Timber Wolf" Battalion and the 114th "Brock's Rangers" Battalion, were composed of roughly 50 per cent Indigenous Canadians. These unique formations were raised, trained and administered by commanders who respected, and were active prior to the war, in Native communities, culture, politics and military affairs. Both Lieutenant-Colonel Glen Campbell and his counterpart, Lieutenant-Colonel Andrew T. Thompson, fostered the Native component and heritage of their respective battalions and took pride in leading their soldiers into war. Both ensured, in the words of Native veteran Daniel Pelletier, that "there was no discrimination 'over there' and we were treated good."[96] Both Andrew Thorburn Thompson, soldier and politician, and Glenlyon Campbell, frontiersman, politician and soldier, encompass the virtues, legacy and foundation of a dignified Canadian past.

1 Relative in the sense that neither the Crimean War (1853-1856) nor the Franco-Prussian War (1870-1871) were fought by numerous allied states nor did the theatres of operations cover vast regions of continental Europe.

2 For the purposes of this chapter, the terms "Native", "Aboriginal", "First Nations" and "Indigenous" will be used interchangeably to refer to the original peoples of North America (Canada). The term "Indian" will only be used as it appears in direct quotations or in government acts or policies.

3 Hew Strachan, *The First World War Volume I: To Arms* (Oxford: University Press, 2001), p.497. In his 1910 book, *La Force noire*, General Charles Mangin predicted the French could raise 40,000 men from its West African colonies (4% of the total population of 10.65 million). By the end of the war, France had enlisted 200,000 soldiers from these same colonies.

4 Shelby Cullom Davis, *Reservoirs of Men: A History of the Black Troops of French West Africa* (Westport: Negro Universities Press, 1970), p.143.

5 Christopher Pugsley, *Te Hokowhitu A Tu: The Maori Pioneer Battalion in the First World War* (Auckland: Reed Publishing Ltd., 2006), pp. 34-36.

6 House of Lords Record Office, London, UK, Andrew Bonar Law Papers, BL/55/16—Memorandum: Colonial Office to Governors-General and Administrators of British Dominions, Colonies and Protectorates, 8 October 1915.

7 Duncan Campbell Scott, *Report of the Superintendent General Indian Affairs, Sessional, Paper No. 27—Indians and the Great War* (Ottawa: Department of Indian Affairs, 1919), p.13; Colonial Office, *The Colonial List, 1915* (London: Waterlow & Sons Ltd., 1914, 1915), p.638. The definitive number of Natives to serve in the First World War is not known. Many "Status Indians" were not recorded as such upon enlistment and records rarely included Métis, Inuit or those Natives from the Territories and Newfoundland/Labrador (which was a distinct British Colony not entering Canadian Confederation until 1949).

8 "Canadian Indians and World War One," *Saskatchewan Indian Federated College Journal* 1, 1 (1984): 67.

9 LAC RG24, Vol. 1221, File HQ 593-1-7. Letter 22 October 1915 from Deputy Minister of Militia Sir Eugene Fiset to W.R. Brown, Indian Agent Port Arthur, Ontario.

10 As quoted in L. James Dempsey, *Warriors of the King: Prairie Indians in World War I* (Regina: University of Regina Press, 1999), pp.38-39.

11 As quoted in Ibid, p.39. NOTE: The assumption in the language is that Natives would only be drafted to fight in the advent of war with the United States.

12 LAC RG10, Vol. 6762, File 452-2-1. Letter of August 1914 from Chief F.M. Jacobs, Sarnia Chippewa Reserve to Duncan Campbell Scott.

13 Robert S. Allen, *His Majesty's Indian Allies: British Indian Policy in the Defence of Canada,* 1774-1815 (Toronto: Dundurn Press, 1992), p.193; Royal Commission on Aboriginal Peoples, *Treaty Making in the Spirit of Co-existence* (Canada: Communication Group Publishing, 1995),

pp.24-25. Allen refers to it as the *Magna Charta* of Native Rights in Canada. It still seen by Canadian Natives as their equivalent to the Canadian Charter of Rights and Freedoms, having been enshrined in Section 25 of the charter itself within the Constitution Act of 1982. It is a little known fact that the British rescinded the Proclamation in 1768 due to American frontiersmen ignoring the order not to settle westward on Native lands. With fear of a revolution, which inevitably came, the British under William Johnson negotiated a treaty with the Iroquois for Ohio—The Treaty of Fort Stanwix, November 1768.

14 Mike Mountain Horse, *My People, The Bloods* (Calgary: Glenbow-Alberta Institute and the Blood Tribal Council, 1979), pp.139, 144.

15 LAC RG10, Vol. 2837, Reel C-11284, File 171,340. Letter of 11 May 1898 from W. Hamilton Merritt to Deputy Minister Indian Affairs. Merritt was the grandson of William Hamilton Merritt, who had been an early Canadian politician and was the driving force behind the building of the Welland Canal. The former was made Honorary Chief by the Cayuga in 1886. He had served in the Boer War as second in command of the 2nd Regiment, Canadian Mounted Rifles.

16 Mountain Horse, *My People, The Bloods*, pp.139-140; Dempsey, *Warriors of the King: Prairie Indians in World War I*, p.18.

17 Thomas A. Britten, *American Indians in World War I: At Home and at War* (Albuquerque: University of New Mexico Press, 1997), pp.102-103; Bruce White, "The American Army and the Indian," contained in N.F. Dreisziger ed., *Ethnic Armies: Polyethnic Armed Forces from the Time of the Hapsburgs to the Age of the Superpowers* (Waterloo: Wilfrid Laurier University Press), pp.78-79.

18 LAC RG10, Vol. 2837, Reel C-11284, Files 171, 340. Composition 37th Haldimand Rifles, Canadian Militia.

19 LAC RG10, Vol. 2837, Reel C-11284, File 171,340. Letter of 14 February 1896 from Chief Josiah Hill, Six Nations Reserve to Captain D.E. Cameron, Indian Affairs Superintendent, Brantford, Ontario.

20 Captain D.E. Cameron, "The Royal Six Nations Regiment," *The Indian Magazine Brantford*, III, 4 (January 1896): 1-4.

21 LAC RG10, Vol. 2837, Reel C-11284, File 171, 340. Letter of 26 Feb. 1896—D.C. Scott, Superintendent of Indian Affairs to Colonel Panet, Deputy Minister of Militia.

22 LAC RG10-Vol. 2991-Reel C-11307, File 215, 977. Letter of 10 Nov. 1899—Six Nation Council Chiefs (Josiah Hill) to Queen Victoria.

23 LAC RG10-Vol. 2991-Reel C-11307, File 215, 977. Various letters pertaining to Native troop contributions.

24 LAC RG10-Vol. 2991-Reel C-11307, File 215, 977. Letter of 13 February 1900—Joseph Chamberlain, Colonial Office to Governor-General Lord Minto and Captain D.E. Cameron, Superintendent Brantford/Six Nations; LAC RG10-Vol. 2991-Reel C-11307, File 215, 977. Indian Affairs Notice of 22 April 1900 to all Reserve and Area Superintendents.

25 John Moses, *A Sketch Account of Aboriginal Peoples in the Canadian Military* (Ottawa: Department of National Defence, 2004), p.61.

26 LAC RG38, A-1-a, Vol. 76. Private George McLean—Personal Records South Africa 1899-1903; Janice Summerby, *Native Soldiers, Foreign Battlefields* (Ottawa: Veterans Affairs, 2005), pp.14-15.

27 LAC RG24,Vol. 1221, File 593-1-7. Departmental Letters: Ministry of Militia and Ministry of Indian Affairs, 1914-1915.

28 LAC RG150, Box 2648 – 15-25. Personal Records for eight soldiers last name Doxtator, 1914-1919; LAC RG150, Box 5135 – 5-7. Personal Records for three soldiers last name Kick, 1914-1919; LAC RG150, Box 5885 – 54-55. Personal Records for two soldiers last name Maness, 1914-1919; LAC RG150, Box 5910 – 1-27. Personal Records for 27 soldiers last name Maracle, 1914-1919; LAC RG150, Box 7241 – 43-47. Personal Records for six soldiers last name Nawash, 1914-1919; LAC RG150, Box 6307 –1-15. Personal Records for 15 soldiers last name Montour, 1914-1919.These names were picked as they are primarily only Native surnames. Of these names checked 50% listed "Indian" under complexion with another 45-50% listing dark for complexion, black for hair and brown for eye colour – all Native traits.

29 Cook, *At the Sharp End: Canadians Fighting the Great War 1914-1916*, Vol. 1, p.71. Corporal Pegahmagabow was one of only 39 Canadians to receive the Military Medal three times during the war. He recorded an unconfirmed 378 kills. After the war he became Chief of the Parry Island Ojibwa Band and an influential figure in Native rights activism. See: Adrian Hayes, *Pegahmagabow: Legendary Warrior, Forgotten Hero* (Sault Ste. Marie: Fox Meadow Creations, 2006).

30 Native Canadians killed during the 2nd Battle of Ypres: Lieutenants Albert Mountain Horse (Alberta, Blood), Cameron D. Brant (Ontario, Mohawk) and Privates Albert Cook (Manitoba, Ojibwa) and Angus LaForce (Quebec, Mohawk). Lieutenant Brant from the Six Nations Reserve was the great-grandson of Iroquois Chief (and British Colonel) Joseph Brant, veteran of the Seven Years' War, the American Revolution, Pontiac's Rebellion and founder of the Six Nations Reserve. Private LaForce from Kahnawake, Quebec, was likely the first Native Canadian killed in the First World War during the first gas attack on 22 April 1915.

31 Dempsey, *Warriors of the King: Prairie Indians in World War I,* pp.19-24.

32 Moses, *A Sketch Account of Aboriginal Peoples in the Canadian Military*, p.62; Gaffen, *Forgotten Soldiers*, p.20. In addition, the Six Nations Iroquois believed that they were a sovereign nation and allies of England and that any request should come from the King of England not from Canada.

33 Dempsey, *Warriors of the King: Prairie Indians in World War I*, p.21.

34 Ibid., p.21.

35 LAC RG24, Vol. 1221, File HQ 593-1-7. Letter of 25 October from Indian Affairs Superintendent A.J. Duncan to Deputy Minister of Militia, Sir Eugene Fiset.

36 LAC RG24 Vol. 4380, File 34-7-89, Vol. 1. Letter of 6 December 1915 from Minister of Militia Sir Sam Hughes to Colonel E.S. Baxter, Commanding Officer 37th Haldimand Rifles, Cayuga, Ontario.

37 LAC RG150, 1992-93/166, Box 518-9—Edwy Sutherland Baxter, First World War Service Records; LAC RG24 Vol. 4380, File 34-7-89, Vol. 1. WWI Organization—114th Overseas Battalion, Haldimand County.

38 LAC RG150, 1992-93/166, 9622—First World War Service Records LCol. Andrew T. Thompson. Five generations of Thompsons (of Scottish ancestry) lived at the Ruthven Park estate outside Cayuga from 1845-1993, including Andrew Thompson. The 1,500 acre Ruthven Park is now a Canadian National Historic Site.

39 Parliament of Canada, Senators and Members—Andrew Thorburn Thompson. <http://www.parl.gc.ca>.

40 Scott, *Report of the Superintendent General Indian Affairs, Sessional, Paper No. 27—Indians and the Great War*, p.7.

41 Thompson as quoted in Marilynn Havelka, "Christmas at Ruthven Park," *Riversong: Lower Grand River Land Trust Newsletter* 8, 4 (December 2001): Chapter 8.

42 LAC RG24 Vol. 4380, File 34-7-89, Vol. 1. WWI Organization—114th Overseas Battalion, Haldimand County.

43 LAC RG24 Vol. 4380, File 34-7-89, Vol. 1. Letter of 31 May 1916 from Duncan Campbell Scott to Brigadier-General W.A. Logie.

44 LAC RG150-1, Vol. 95. Various Documents relating the 114th Overseas Battalion; James W. St.G. Walker, "Race and Recruitment in World War I: Enlistment of Visible Minorities in the Canadian Expeditionary Force," *Canadian Historical Review* LXX, 1 (1989): 13.

45 J.R. Miller, *Skyscrapers Hide the Heavens* (Toronto: University Press, 2000), pp.65-68.

46 Walker, "Race and Recruitment in World War I: Enlistment of Visible Minorities in the Canadian Expeditionary Force," pp.9, 13.

47 Ibid., p.14. In July 1916, when Colonel Mewburn called for a report on Natives enlisted in Military District 2, headquartered in Toronto, the 114th had 348 Natives, including 5 officers, while 211 Natives were arrayed across 15 other units.

48 LAC RG9-II-B-10, Vols. 31, 38. Nominal Rolls, 114th Overseas Battalion; LAC RG24-C-1-a, Vol. 1562, File 1. Pay and Pay Sheets, 114th Overseas Battalion.

49 LAC RG24 Vol. 1562, File HQ 683-173-2. Letter of 25 March 1916 from Lieutenant-Colonel A. Thompson to Adjutant-General, Ministry of Militia.

50 Summerby, *Native Soldiers, Foreign Battlefields*, p.7; LAC RG24 Vol. 1562, File HQ 683-173-2. Letter of 25 March 1916 from LCol. A. Thompson to Adjutant-General, Ministry of Militia.

51 LAC RG24 Vol. 1562, File HQ 683-173-2. Inspection Report, 114th Overseas Battalion, 17 October 1916.

52 Robert was a descendant of the Campbell Clan, members of which were the perpetrators of the Glencoe Massacre (Scotland) on 13 February 1692 during the era of the Glorious Revolution and Jacobitism. Seventy-seven members of the MacDonald Clan, including women and children, were killed (or died as a result of the massacre) by Captain Robert Campbell and his followers.

53 Grant MacEwan, Fifty Mighty Men (Saskatoon: Western Producer, 1958), pp.116-119; Kenneth Stephen Coates, *Dictionary of Canadian Biography—Robert Campbell.* <http://www.biographi.ca>.

54 Glen also contracted typhoid at this time.

55 Desmond Morton, *Dictionary of Canadian Biography—Glenlyon Archibald Campbell;* MacEwan, *Fifty Mighty Men*, pp.116-118.

56 Charles Arkoll Boulton, *Reminiscences of the North-west Rebellions: with a record of the raising of Her Majesty's 100th regiment in Canada, and a chapter on Canadian social and political life* (Toronto: Grip Printing and Publishing Co., 1886), Chapter XIV-Batoche Captured; LAC RG150, 1992-93/166, 1434—First World War Service Records Lt. Col. Glen Campbell.

57 Coates, *Dictionary of Canadian Biography—Robert Campbell; Bruce Vandervort, Indian Wars of Mexico, Canada and the United States 1812-1900* (New York: Routledge, 2006), pp.161-191.

58 According to the town of Gilbert Plains, Manitoba, in 1884, Campbell rode his horse over the Riding Mountains into what is now known as the Gilbert Plains, finding only one man, Métis Gilbert Ross and his wife, living in a small cabin. Campbell traded his horse for the cabin and moved in. See: O.E.A. Brown, *Settlers of the Plains* (Gilbert Plains: The Maple Leaf Press, 1953).

59 MacEwan, *Fifty Might Men*, pp. 120-121; Morton, *Dictionary of Canadian Biography—Glenlyon Archibald Campbell*; Parliament of Canada, Senators and Members—Glenlyon Archibald Campbell. Sifton was the MP for Brandon, Manitoba. He was Minister of the Interior and Superintendant General for Indian Affairs in Laurier's cabinet from 1896-1905. He was largely responsible for the influx of eastern Europeans (some 3 million) to western Canada at the turn of the century. He was knighted by King George V on 1 January 1915.

60 Morton, *Dictionary of Canadian Biography—Glenlyon Archibald Campbell.*

61 *The New York Times*, "Branded as Liar in Ottawa House," 18 March 1911.

62 MacEwan, *Fifty Mighty Men*, p.121. Another anecdote reportedly has Campbell re-enacting his account of riding a bull moose for members of the House of Commons.

63 LAC RG10, Reel C-10204, Vol. 4063, File 402890—Correspondence Regarding the Appointment of Glen Campbell as Chief Inspector of Agencies, Reserves and Inspectorates in Manitoba, Saskatchewan, Alberta and the Northwest Territories. Also Subsequent Work while on Loan to Department of Militia & Defence and Death as Lieutenant-Colonel in France. Saulteaux are a branch of the Ojibwa Nation within the Algonquian language grouping. They are also referred to as Anishinaabe. Harriet Campbell (nee Burns) died on 17 May 1910 at the age of 44.

64 LAC RG150, 1992-93/166, 1434—First World War Service Records Lt. Col. Glen Campbell.

65 L. James Dempsey asserts that perhaps Campbell's initial motivation to recruit underage Natives was simply to increase his battalion's strength. Campbell could also have been using his former position of power within the Ministry of Indian Affairs and the Ministry itself to undermine or intimidate Native parents, Chiefs and Band Councils into agreeing to this policy.

66 Dempsey, *Warriors of the King: Prairie Indians in World War I*, pp.24-25; Colonial Office, Colonial List, 1915, p.640. At this time in Manitoba, 795 Native boys were attending various "Indian Schools and Institutions."

67 Morton, *Dictionary of Canadian Biography—Glenlyon Campbell.*

68 Colonial Office, Colonial List, 1915, p.640. Statistics for the rest of the Canadian Provinces are as follows: P.E.I.—61%; Nova Scotia—62%; New Brunswick—50%; Quebec—26% English, 48% French; Saskatchewan—22%; Alberta—15%; B.C.—33%. Given that most of the Native members of the 107th would have come from Manitoba, Saskatchewan and Alberta, it is likely that the majority spoke little or no English.

69 LAC RG9 Series III-D-3 Vol. 5010, Reel T-10859, File 725. War Diaries—107th Canadian Pioneer Battalion; Dempsey, *Warriors of the King: Prairie Indians in World War I*, p. 25; Gaffen, *Forgotten Soldiers*, pp.23-24.

70 Steven A. Bell, "The 107th 'Timber Wolf' Battalion at Hill 70," *Canadian Military History* 5, 1 (Spring 1996): 78.

71 Interview by Author with Glenlyon Campbell (great-grandson of Lieutenant-Colonel Glen Campbell), 15 January 2008.

72 LAC RG9, III-D-3, Vol. 5010, Reel T-10859, File 725—War Diaries 107th Pioneer Battalion; LAC RG24 Vol. 1562, File HQ 683-173-2. Inspection Report, 114th Overseas Battalion, 17 October 1916.

73 Interview by Author with Glenlyon Campbell (great-grandson of Lieutenant-Colonel Glen Campbell), 15 January 2008; LAC RG150, 1992-93/166, 1445-26—First World War Service Records Lt. John Robert Campbell; LAC RG150, 1992-93/166, 1434—First World War Service Records LCol. Glen Campbell. Glen Campbell married Florence Wesley in 1916 with whom he had one son (21 Oct. 1917). His first wife, Harriet (nee Burns) Campbell, died on 17 May 1910 at the age of 44. His 4 children (3 daughters and 1 son) with Harriet persuaded him to exclude his new wife from his will.

74 LAC RG9-II-B-10, Vols. 31, 38. Nominal Rolls, 114th Overseas Battalion; LAC RG24-C-1-a, Vol. 1562, File 1. Pay and Pay Sheets, 114th Overseas Battalion; LAC RG150, 1992-93/166, 1434—First World War Service Records LCol. Glen Campbell.

75 Andrew T. Thompson, *Report by Col. Andrew T. Thompson Commissioned to Investigate and Enquire into the Affairs of the Six Nations Indians, 1923* (Ottawa: F.A. Acland, King's Printer, 1924), pp.12-13; Donald B. Smith, *Dictionary of Canadian Biography—Deskaheh (Levi General)*; Ronald Niezen, "Recognizing Indigenism: Canadian Unity and the International Movement of Indigenous Peoples," *Comparative Studies in Society and History* 42, 1 (January 2000): 123-124. During this time, traditional Longhouse Chief Deskaheh travelled to Geneva to lobby the League of Nations for international recognition of the Six Nations Iroquois Confederacy as an independent Nation, under article 17 of the League's covenant.

76 Andrew Thorburn Thompson, *The Odyssey of the Gentlemen Adventurers Travelling into the Brazils* (Ottawa: Privately published, 1928).

77 LAC RG150, 1992-93/166, 9622—First World War Service Records LCol. Andrew T. Thompson.

78 LAC RG9-II-B-10,Vols. 31, 38. Nominal Rolls, 114th Overseas Battalion; LAC RG24-C-1-a, Vol. 1562, File 1. Pay and Pay Sheets, 114th Overseas Battalion; Scott, Report of the Superintendent General Indian Affairs, Sessional, Paper No. 27—Indians and the Great War, pp.16-17.

79 Gaffen, *Forgotten Soldiers*, p.24. In total, at least 14 Native Canadians were commissioned officers during the First World War.

80 Bell, "The 107th 'Timber Wolf' Battalion at Hill 70," pp.73-74.

81 LAC RG9, III-D-3, Vol. 5010, Reel T-10859, File 725—War Diaries 107th Pioneer Battalion; Morton, *Dictionary of Canadian Biography—Glenlyon Campbell.*

82 LAC RG9, III-D-3, Vol. 5010, Reel T-10859, File 725—War Diaries 107th Pioneer Battalion.

83 LAC RG9, III-D-3, Vol. 5010, Reel T-10859, File 725—War Diaries 107th Pioneer Battalion; Bell, "The 107th 'Timber Wolf' Battalion at Hill 70," pp.74-76.

84 LAC RG9, III-D-3, Vol. 5010, Reel T-10859, File 725—War Diaries 107th Pioneer Battalion; Scott, *Report of the Superintendent General Indian Affairs, Sessional, Paper No. 27—Indians and the Great War*, p.15. The total number of Canadian casualties at Hill 70, from 15-25 August, was 9,198. Six Victoria Crosses were awarded to Canadian soldiers during this action.

85 LAC RG9, III-D-3, Vol. 5010, Reel T-10859, File 725—War Diaries 107th Pioneer Battalion.

86 Ibid.

87 Scott, *Report of the Superintendent General Indian Affairs, Sessional, Paper No. 27—Indians and the Great War*, p.15.

88 LAC RG9, III-D-3, Vol. 5010, Reel T-10859, File 725—War Diaries 107th Pioneer Battalion; Bell, "The 107th 'Timber Wolf' Battalion at Hill 70," pp.77-78. Longboat, an Onondaga Iroquois, had won the 1907 Boston Marathon in a record time of 2:24:24 and also participated in the 1908 London Olympics. Keeper, also a professional runner, participated in the 1912 Stockholm Olympics.

89 LAC RG150, 1992-93/166, 1434—First World War Service Records LCol. Glen Campbell; LAC RG9, III-D-3, Vol. 5010, Reel T-10859, File 725—War Diaries 107th Pioneer Battalion; LAC RG9 III-D-3, Vol. 4990, Reel T-10825, File 641—War Diaries 1st Brigade Canadian Engineers; Commonwealth War Graves Commission, *Casualty Details—LCol. Glenlyon Campbell,* <http://www.cwgc.org>.

90 LAC RG9 III-D-3 Vol. 4946, Reel T-10753-10754, File 470. War Diaries—160th Canadian Infantry Battalion; LAC RG9 III-D-3 Vol. 4941, Reel T-10748, File 442-443. War Diaries—52nd Canadian Infantry Battalion; Dempsey, *Warriors of the King: Prairie Indians in World War I*, p.23; , The (Regina, Saskatchewan) Leader, "Indians are doing their bit in the Great War," 18 November 1916.

91 See: Pugsley, *Te Hokowhitu A Tu: The Maori Pioneer Battalion in the First World War;* James Cowan, *The Maoris in the Great War* (Auckland: Whitcombe & Tombs Limited, 1926); P.S. O'Connor, "The Recruitment of Maori Soldiers, 1914-1918," *Political Science* 19, 2 (December 1967): 48-83.

92 Major-General Gwatkin as quoted in Dempsey, *Warriors of the King: Prairie Indians in World War I*, p.22.

93 Lisa Sattenspiel and Dawn Herring, "Structural Epidemic Models and the Spread of Influenza in the Central Canadian Sub-Arctic," *Human Biology* (February 1998): 3.

94 W.M. Graham as quoted in ibid., p.24; Britten, *American Indians in World War I: At Home and at War*, pp.74-75; Michael L. Tate, "From Scout to Doughboy: The National Debate over Integrating American Indians into the Military, 1891-1918," *The Western Historical Quarterly*, 17, 4 (October 1986): 430. The United States' policy towards their Native soldiers was very similar to that of Canada. They maintained a policy of integration into white units. However, like Canada, they did have units that were made up primarily of Natives such as 158th and 358th Infantry Regiments. The 142nd Infantry Regiment contained over 600 Cherokee, Choctaw, Chickasaw, Creek and Seminole Natives.

95 Canadian War Museum, Ottawa. The first contingent of the CEF contained only one company of French Canadians. The 22nd Battalion "Van Doos" was formed in early 1915. A mere 13 of the 258 Infantry Battalions raised throughout the course of the war were raised in Quebec. Most could not retain their required strengths and all were broken up to reinforce the 22nd or were scattered into English speaking battalions. In addition, nearly half of the French Canadian volunteers (15,000) came from the provinces of New Brunswick, Nova Scotia and Ontario. Sir Sam Hughes was fired in November 1916.

96 Dempsey, *Warriors of the King: Prairie Indians in World War I*, p.50; Moses, *A Sketch Account of Aboriginal Peoples in the Canadian Military*, p.1. According to Moses, as of 2003, there were 1,200 First Nations, Inuit and Métis Canadians serving in the Canadian Forces, representing 640 bands and 55 linguistic dialects from 11 linguistic families.

CHAPTER 8

Creating Combat Leaders in the Canadian Corps

The Experience of Lieutenant-Colonel Agar Adamson

TOD STRICKLAND

It is a paradox, disconcerting to the historian, that complete successes make less history (in the case of small units, at least) than does failure or partial achievement.

-Ralph Hodder-Williams [1]

The Princess Patricia's Canadian Light Infantry played a relatively small part in the Canadian attack on Vimy Ridge within the context of the larger Canadian Corps.[2] The Regiment was successful in taking its objectives with comparatively low casualties but it was only one of forty-eight Canadian infantry battalions who took part in the attack on the German positions dominating the Douai plain.[3] Placed in the centre position of the 7th Canadian Infantry Brigade,[4] their contribution (along with that of 7 CIB and the entire 3rd Canadian Division) to the endeavour's success has largely been overlooked. Studies generally focus on the more dramatic events: the costly efforts to take Hill 145 and "the Pimple," or the length of the advance in the First and Second Canadian Division sectors. Viewed in contrast to the struggles on other parts of Vimy Ridge, the Patricia's were quite successful in their attack, with almost every aspect unfolding according to plan.

Success in battle does not just happen, as was clearly evident during the First World War. Whether at the strategic, operational or tactical levels, victory was almost always the product of comprehensive planning, thorough preparations and strong combat leadership. Without capable commanders, the best equipped and trained troops were unlikely to be successful in the accomplishment of their tasks. This truism remains valid today.

Combat leaders were not grown overnight. Instead they were, and are, the product of time-consuming education and experience. Effective COs were the result of deliberate formal and informal efforts by higher commanders to train their subordinates in the art and science of war.[5] In the latter half of the First World War, this was a sphere in which the Canadian Corps excelled.

However, much of the Canadian historical literature has neglected the study of unit COs.[6] Rather than detail the reasons for success at the Battalion level, Canadian First World War history is replete with examples at the divisional and corps level; perhaps because this was the first time that Canadian officers attained this level of command. Similarly, much has been written on the entirety of actions that were undertaken by Canada on 9 April 1917. It is not an overstatement to argue that much has been written about *what* was done by the Canadian Corps on that Easter Sunday, but precious little has been done to describe *how* unit COs were able to achieve their goals. Examining the actions, preparations and training of unit-level commanders affords insight into both the processes of command and the leadership traits that were evident in infantry battalion COs in mid-1917.

Lieutenant-Colonel Agar Adamson was not extraordinary. Older than most of his peers, blind in one eye and possessing previous experience in the Boer War, he was the CO of the PPCLI for eighteen months. This included both the build-up and preparations for the attack on Vimy Ridge, as well as the actions which followed into the spring of 1918. This examination of Adamson's command is focused on the period from the moment he took command of the unit in October 1916, until the Regiment had completed its assigned tasks atop the Ridge on 11 April 1917. The intent is to demonstrate how COs were readied for the critical role they played in ensuring success on the battlefield during the First World War, demonstrating that the development of tactical excellence was both a formal and an informal process undertaken by commanders at all levels – educating their subordinates while themselves being trained.

A short examination of the Agar Adamson's life before the war will highlight the very Canadian nature of the man. A description of the Regiment which formed the operating parameters for Adamson as he advanced in both rank and experience provides a contextual understanding of the working environment and the formative experiences he had prior to being appointed to command the Regiment. A study of the expectations that were made of commanding officers and how they were trained to fill their role will show that Adamson was not unique in his position, training or experiences.

Prior to delving into the study however, it is worthwhile to understand that there are distinct differences between the terms "leadership" and "command" and to more closely examine how COs are prepared for their roles. Currently the Canadian Forces define leadership as "directly or indirectly influencing others, by means of formal authority or personal attributes, to act in accordance with one's intent or a shared purpose."[7] As such it is related to "command", which may be considered a complementary concept. In short, to lead is to influence another person to a common goal. Hence, when a person is referred to as a leader, he is filling an influence role.

By contrast, command is "the purposeful exercise of authority – over structures, resources people, and activities."[8] The difference, though subtle, is important. When a commander commands he is ultimately relying on the authority vested in his position. What should be evident is that a good commander should also be a leader – harnessing the use of influence to achieve his goals, as opposed to strictly relying on the authority that he has been given.[9]

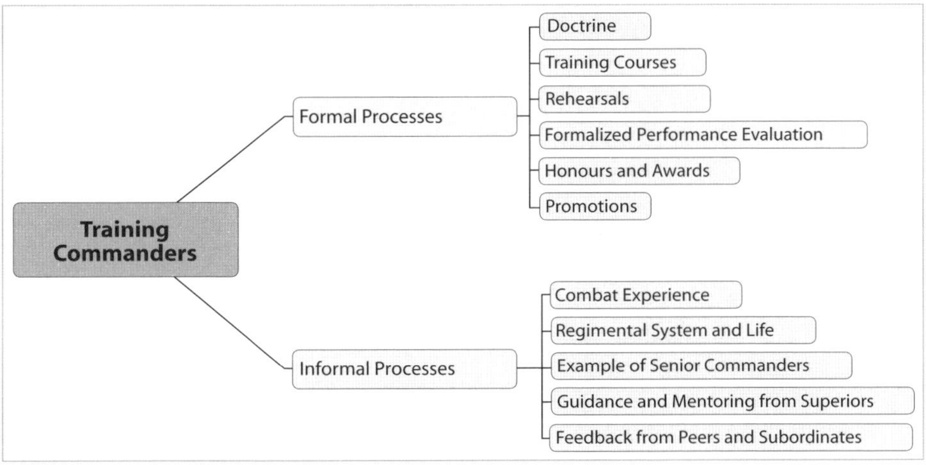

FIGURE 1 – TRAINING COMMANDERS

There are both formal and informal processes at work in training commanders, as can be seen above. Formally, professional military organizations instruct their leaders by exposing them to doctrine (both in written and other forms), as well as by mandating training for them. Actions are rehearsed in front of superior officers to confirm understanding, and feedback sessions are used to clearly articulate how one's performance is perceived. Particularly strong efforts are rewarded by the institution, using both promotions and by the giving of honours or awards. This sets benchmarks for performance that are readily apparent to all who would aspire to command. It also systemically reinforces when a commander's actions were in line with what his superiors desired.

Informally, other processes work to educate commanders. Experience on operations is particularly formative; the enculturation of officers that accompanies belonging to a regiment is equally important. It is in this informal schoolhouse that one learns what is expected, the relationship between the commander and his subordinates, and the intricacies and nuances of both tactics and leading soldiers. Similarly, commanders are prepared by the example of their superiors. It is an accepted truism that soldiers copy whatever example is set for them, hence the importance placed upon setting the example within contemporary military leadership doctrines.[10]

This is further enhanced by guidance and mentoring from one's superiors, both in the workplace and in social settings. Lastly, and perhaps of more importance, is the feedback one gets from both peers and subordinates. The unvarnished opinion given by one's troops, in particular, offers clear and unequivocal evidence as to how one is performing. A good commander should at least consider how he or she is being perceived in carrying out his or her duties. These processes are readily evident in the life of Agar Adamson.

AGAR STEWART ALLEN MASTERTON ADAMSON

On Christmas Day 1865, Agar was born in Ottawa "into the old Upper Canadian gentry."[11] Little is noted from his early years, aside from the fact that he grew up in what would become the national capital, where his father had become a clerk in the Senate.[12] Following some academic dabbling in Cambridge, courtesy of a wealthy uncle, and after considering joining the priesthood, he returned to

Ottawa.[13] Here in 1890, at nearly twenty-five years of age, he followed in his father's footsteps and became a junior clerk in the Senate.

The first experience Adamson had with the military came in the form of an appointment into the Governor General's Foot Guards in 1893. The newly gazetted lieutenant quickly evolved into a man about town, becoming linked socially to the Governor General and his wife. As observed by author Sandra Gwyn, Adamson was "a favourite at the slightly louche vice-regal court" which in turn afforded him opportunities he would never have seen as a clerk.[14] Two years later in 1895, while returning from a military training exercise, he stopped in Toronto to attend a dinner party; also present was Miss Mabel Cawthra. In 1899, after a "stormy" courtship, the pair were married.[15]

Shortly before their marriage, war broke out in South Africa between the British and the Boers. Almost immediately, the Laurier government announced that it was sending a Canadian contingent. Lieutenant Adamson tried to get overseas in January 1900 by joining the Special Service Regiment which had been raised to relieve the British garrison at Halifax. Three months later, an offer came to join the third Canadian contingent then being raised (and paid for) by Donald Smith – the Lord Strathcona.[16] Adamson seized the opportunity and soon left to join his new Regiment in South Africa.

The experiences that he gained with the Lord Strathcona's Horse were seminal in his formation as a military officer. In the words of Sandra Gwyn:

> Adamson's experiences on the veld [sic] were transformational. He served with distinction, winning a Mention in Dispatches. He had a natural gift for command, and an innate respect for his men.[17]

These thoughts are echoed in the recent work of Craig Leslie Mantle, who attributes much of Adamson's later success with the PPCLI to his professional schooling in warfare in South Africa.[18] In Adamson's Boer War experience we see the beginning of many of the traits that would later typify his style of command and leadership in the Patricias: a sense of respect for his soldiers and an abiding interest in their welfare, a strong sense of discipline, and use of competence as a guiding principle in determining who his subordinate leaders would be.[19] After becoming seriously ill, which was not uncommon among the British Empire's soldiers in South Africa, he departed the theatre in November 1900 and returned to Canada.[20]

Adamson does not appear to have been happy back in Ontario and out of the action. Following a significant degree of lobbying, he became a junior captain in the 6th Battalion, Canadian Mounted Rifles. Boarding a troopship in mid-May 1902, he and his men arrived in South Africa after the war ended. No longer required for use against the Boers, the unit returned to Canada. Adamson's active military service, for the moment, came to a close.[21] Nearing forty years old, he left the Senate in 1903 and moved to Toronto to work as the titular head of a British decorating firm.[22]

When one looks at the man himself, it immediately becomes apparent that he was an engaging extrovert. This is best shown in the language used by his principal biographer, Sandra Gwyn. As she writes:

> The most memorable of his attributes was his charm... [it] encompassed a sense of humour that was both prankish and ironic, a disarming candour, and... an easy bonhomie... He was a man's man who got on famously with women...[23]

Further evidence can be seen in the extensive network of social connections that he maintained throughout his life. The Governor General and his wife, the General Officer Commanding the Canadian Militia, and Lord Strathcona, for example, were all known to Agar as a young man; this would not have been possible, and the connections would not have been maintained, were it not for a high level of innate confidence in his own abilities.

Adamson's Boer War and early military experiences were important for three separate reasons. First, they put him firmly in the average when it comes to previous military experience among the men who commanded Canadian infantry battalions during the First World War. As noted by historian Patrick Brennan, of the 179 men who commanded battalions in France for an extended period of time, 160 had the benefit of "pre-war military experience" before taking command. A further 24 of them (roughly one in eight) had seen action on the veldt.[24]

As such, this supports the assertion that Adamson was not extraordinary in later being a CO.

Second, his time with the Strathconas during the Boer War formed the basis of the combat experience that he would rely on at the outset of the First World War. This was a critical piece in the informal aspects of Adamson's training. It was here that he first saw how a CO both acted and commanded; an example of which he was particularly critical.[25] Additionally, it was here that he first experienced being under fire, made tactical errors and gained confidence in his ability to perform acceptably while being engaged by an enemy. This was reinforced by his winning of a Mention in Dispatches.

A last point to note is that his experiences in South Africa also served to expand the social network that Adamson appears to have cultivated throughout his life. Two personalities in particular stand out. The first is Julian Byng, who commanded the Canadian Corps at Vimy Ridge. Agar appears to have worked for Byng directly while attached to the South African Light Horse when Byng was the CO of that unit.[26] The other is Adamson's future Brigade Commander at Vimy Ridge, A.C. "Batty Mac" Macdonell. The two men met when they both served as officers in Lord Strathcona's Horse, where Macdonell was a cavalry officer in South Africa.[27]

The outbreak of the First World War shook Adamson out of the relatively comfortable life he was leading, and, like many of his day, saw him volunteer to join the Canadians who were about to head overseas. The regiment that he chose to join was the PPCLI.[28]

THE "PATRICIAS"

If Adamson was in many ways average, the PPCLI was not. Formed by Montreal businessmen Hamilton Gault (who was himself a Boer War veteran)[29] as war broke out on the European continent, the Regiment was the last privately raised unit within the Commonwealth. It was distinct from the outset primarily because it drew its recruits from across Canada. Only the Royal Canadian Regiment could claim that it too drew is members from across the continent. Most militia units, by contrast, were raised from a given geographical area within the various provinces of Canada.

Additionally, veterans formed the vast majority of the Regiment's recruits. This gave it a truly unique status among the Canadian military establishment; no other unit was similarly manned. As Ralph Hodder-Williams records, 1100 men, out of 3000 volunteers, were selected by Lieutenant-Colonel Farquhar (formerly of the Coldstream Guards and then military secretary to the Governor General) and his small staff.[30] Of the 1100 men who formed the original Regiment, 771 had been granted medals or decorations, 456 had previously war experience, and 1049 had previous military experience.[31] This was to pay significant dividends as the unit went to war, shortening training time and allowing the unit to become the first of Canada's infantry battalions to see service in France, arriving with the British 27th Division in late-December 1914.[32]

In accordance with the latest British tactical doctrine, the Regiment was organized into four rifle companies as shown in Figure 2. At the time, conventional tactics saw the rifle company as the lowest tactical unit that would move on the battlefield. Platoons and sections existed to facilitate administration, not to fight on their own.[33] The Battalion, as an entity, served a myriad number of purposes. These have been described quite well by historian Ian McCulloch, who wrote:

> The infantry battalion was an administrative as well as a fighting unit, which not only trained and organised its members for combat, but fed, paid and clothed them, arranged for their leave and saw to their health, spiritual health and recreational needs.[34]

Although the Regiment began its life in the form prescribed by British doctrine, the doctrine was not static. Between 1914 and the end of the war the Regiment, and indeed all Canadian and British infantry battalions, evolved. Organizationally, the main arm of the Canadian military fighting on the western front continuously adapted to the realities of trench warfare. Innovations like machine gun sections, trench mortar sections, sniper and bombing sections were all developed in a consistent effort to enhance the combat effectiveness of the infantry. Being part of a British division enabled the Regiment to commence this evolution earlier than any of the other battalions that had arrived in Europe at the same time. It also meant that they would not have to use the Ross rifle in battle, instead being issued the Lee-Enfield in late 1914.

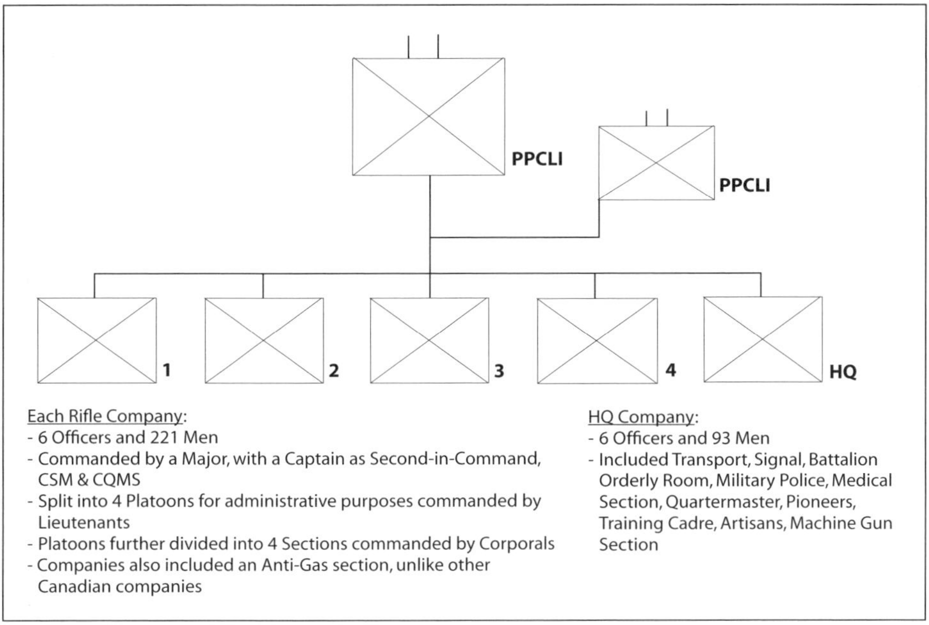

FIGURE 2 – ORGANIZATION OF PPCLI (1914)[35]

After successfully applying to join the Regiment, Adamson was appointed as a captain with No. 3 Company. Disappointingly for him, when the unit departed Britain for France immediately before Christmas 1914, Captain Adamson was left behind to bring forward a draft of reinforcements.[36] However, it was in this capacity that he would have the first of three formative experiences within the Regiment, which would ultimately assist him when he was appointed to lead it in battle.

In early 1915, Captain Adamson witnessed the draft of reinforcements that was allocated to the PPCLI get redirected to a general Canadian replacement pool. Additionally, he found himself cut from his Regiment by the CO of the replacement depot who put him into a pool of surplus officers.[37] After protesting to Colonel James, who commanded the depot, Adamson used his initiative. First he got word to Lieutenant-Colonel Farquhar of James' actions.[38] He then sent a trusted senior Sergeant to meet with Farquhar and explain what had transpired. Within short order, Farquhar resolved the situation and Adamson was again on his way to France with his replacements. His actions, as he recorded to his wife, "were upheld and approved of."[39]

For the Regiment, this incident forced concrete arrangements to be made regarding how its replacements would be handled. For Adamson, there were two main points. The first was he demonstrated that he trusted his NCOs with important tasks. The action he took with Sergeant Cork would have spread around the Regiment, probably in short order, and arguably enhanced his reputation with the soldiers. Second, his actions in using initiative and stepping outside of the chain of command on matters of principle were rewarded. This speaks to the leadership abilities of Lieutenant-Colonel Farquhar; where others might have chastised a subordinate for acting outside of orders, Farquhar's command style instead fostered and rewarded initiative and a certain degree of audacity. As Farquhar wrote, "diffidence be damned."[40]

The second element of Regimental life which assisted Agar as he moved up the chain of command between 1915 and 1916 was the example set by some of the Regiment's first COs. Two of them, Lieutenant-Colonels Farquhar and Buller, were particularly important. After Adamson was appointed to lead No. 2 Company, his contact with the CO would have increased and he would have seen exactly what it was to lead a regiment in combat. One of Adamson's letters clearly shows the kind of leader Farquhar was:

> March 10, 1915...The C.O. took a company out to improve No. 21 at one o'clock in the morning I found him talking to the men at one end of my trench... He had the hardest time of any one in the Regiment – out all night, crawling about from trench to trench, always cheerful and sound...[41]

On 21 March, while Farquhar was moving about the trenches, showing his position to the CO of the unit set to relieve the Patricias, he was struck by a bullet. He died three hours later in a dressing station.[42] His replacement was his Adjutant, Captain H.C. Buller who was subsequently promoted directly to Lieutenant-Colonel.[43] Like his predecessor, Buller was originally a British officer who had left the Canadian Governor General's staff to go to war with the PPCLI. Adamson notes that the similarities extended further. On one occasion Buller demonstrated the same propensity to be out in the trenches at three in the morning (to visit Adamson in his Company HQ),[44] on another he sent Adamson a horse to facilitate his movement back out of the line.[45] This lesson in generosity was one that Adamson himself would emulate.

Six weeks later, after twelve consecutive days in the line, the Patricia's occupied new reserve positions near "Hell-Fire Corner." There, on 5 May, Buller's initial period as CO was cut short when he was hit by artillery shrapnel and blinded in one eye.[46] Command of the unit now fell to the Regiment's founder, Major Hamilton Gault, who had just returned after being previously wounded; Gault, in turn, appointed Adamson second-in-command of the unit. What would become officially known as the Battles of Frezenberg and Bellewaerde Ridge was about to begin.[47] Taking place in the context of the larger Second Battle of Ypres, the next night the Battalion moved back atop Bellewaerde Ridge. On 8 May, the Germans attacked.

The Regiment faced successive heavy German attacks, supported by artillery and machine guns. Twice wounded, the last time quite severely, Gault was forced to turn over command to his number two and Captain Adamson became the CO. The situation was grim: the Regiment was being cut to pieces, they were short of ammunition, and command in the rifle companies was being assumed by NCOs.[48] Adamson, himself now wounded, made sure ammunition was redistributed and then led a counter-attack into a position on the left of his line where the Germans had managed to break in. Around the Patricia's, the Germans were achieving success, but atop Bellewaerde Ridge, the line held.

As their Regimental history recorded, Adamson performed very well. Hodder-Williams wrote:

> Captain Agar Adamson, although painfully wounded, continued to direct the defence with the utmost coolness, heartening the men by cheery words as he moved about distributing ammunition with his unwounded arm.[49]

As the Brigade sent men from other units forward to resupply the Patricias with ammunition, Adamson used them to reinforce his ever-diminishing manpower and prevent the Germans from gaining the ridge.[50] After night had fallen and his casualties had been attended to, he turned command over to one of the few remaining officers, and went to get his wounds dressed.[51] For his actions the first time in command of the unit, he was awarded the Distinguished Service Order (DSO).[52] He would be away from the Regiment recovering from his wounds until September.

The battle and his relatively short initial tenure as CO, were important in Adamson's development as a commander. His performance throughout the day of 8 May had been noteworthy, as his superiors observed and officially recognized through the award of the DSO. He gained credibility and a stronger reputation with the soldiers of the unit. This is a sometimes overlooked facet of leadership, however for a combat leader it is of vital importance. Further, by commanding and leading in some of the worst circumstances imaginable, and being successful while doing so, there can be little doubt that his confidence in his own abilities would have been enhanced. This in contrast to the evidence that he considered the war to be badly mismanaged and that he lacked confidence in the standard of generalship that he had seen while a member of the 27th British Division.[53]

For the next year, newly promoted Major Adamson would be with the Patricias in the field, occupying different positions as circumstances demanded. In late 1915, the Regiment was pulled from the British Army to join the 7 CIB in the 3rd Canadian Division.[54] In June 1916 following the fighting at Sanctuary Wood, he again found himself in command after Lieutenant-Colonel Buller, who had rejoined the Regiment, was killed. Historian Jeffery Williams recorded Hamilton Gault's views on Adamson assuming command. He wrote that "Gault had confidence in Adamson's ability to command, a feeling which the latter did not share." He then recorded the words Gault wrote his friend in an attempt "to bolster his morale":

> You are far too modest in yourself old boy. Please remember that there is non better suited to command the Battalion than yourself and, besides this, there is now no one else to handle the job.[55]

Adamson commanded the Regiment until early August, when Lieutenant-Colonel Pelly was brought back to the unit.[56] For all intents, Adamson's apprenticeship was over. The next time he would command the Regiment, it would be after formally being appointed as opposed to taking over because of battle casualties.

THE EXERCISE OF COMMAND

Before reviewing the actions of the Patricias at Vimy Ridge, it is worth examining exactly what was expected of COs from a doctrinal perspective, as well as the means that were afforded them to effect command. As with tactics, the responsibilities given to COs were not static in nature. They changed as new methods, equipment and techniques were tried to break the physical stalemate caused by opposing continuous entrenchments. These directly affected both what COs were expected to do, and how they were expected to do it.

Outside of their roles on the battlefield, battalion commanders were an important link in the processes of both tactical innovation and education. This was observed by Patrick Brennan and Thomas Leppard who argued that COs were central in the development of new tactics through their provision of "comprehensive and frank" reports to higher headquarters as to what worked and what did not in the unceasing effort to break the stalemate. They also note that COs were the personnel most responsible for the training of their junior officers.[57] It was through the CO that lessons were observed, sent higher, and ultimately disseminated.

On the battlefield, their roles and means of carrying them out were governed by doctrine. British and, by extension, Canadian doctrine was quite detailed in what it expected of battalion commanders. It was also quite clear as to the principles which it expected commanders to use on the battlefield. These were expressed at the start of the war in at least three separate publications.[58] The most relevant to our discussion was *Infantry Training (4-Company Organization)* 1914, published by the British War Office.[59] Full of practical information, it offered guidance on where an effective CO should position himself, and on his ability to influence the battle once it had been joined. It stated:

> The commander's position will, as a rule, be selected so that he can obtain an extensive view. It should be sufficiently central to facilitate the receipt of reports and the issue of orders... battalion commanders... will take post where they can best exercise supervision over their commands, watch the enemy, receive and transmit orders...

> During the fight, the commander of a considerable body of infantry influences the course of the action by means of his original orders and subsequently by the employment of his reserve.[60]

There are several elements of these two passages that stand out. The direction concerning placement on the battlefield remains relevant to the current day. Additionally, the book places heavy emphasis on getting one's original orders correct. This would seem to place less importance on the exercise of initiative when clear orders were unavailable. Lastly, the only means available to a commander to influence the battle was his reserve. By the modern definition, a reserve is a force that is not committed to any tasks on the battlefield. It is therefore free to be used to reinforce success or, in the worst case, to try and prevent mission failure. It does not seem that this was exactly the same in 1914, as the publication details that "stragglers and slightly wounded men should be collected and formed into reserves whenever met with."[61] *Infantry Training* does however stress the importance of the reserve:

> Next to the conception of a sound plan of attack, and the issue of clear and comprehensive orders to the company commanders, the most important duty of a battalion commander is the handling of his local reserve... it is by means of this reserve that he makes his influence felt in action...[62]

The book did have its limitations. The changes that an infantry battalion would undergo over the course of the war, such as the addition of new weapons systems like trench mortars and ever-increasing number of machine guns, were not anticipated.

The publication also includes specific guidance concerning how a commander was expected to lead. Micromanagement, as well as over-emphasis on personal example and the use of personal control over "all portions of the force" were to be avoided. These were viewed as distracting from other important tasks that were expected of a CO, such as "protecting the flanks, meeting counterattacks, reporting to or communicating with the superior commander, and maintaining connection with the artillery and adjoining units."[63]

In 1916, because of the massive number of casualties and problems associated with trench warfare, the role of COs underwent a distinct change. This was commented on by historian M.A. Ramsey, who wrote:

> At the Somme, Haig and his corps commanders refused to allow commanders from battalion upwards to participate in major attacks, in order to ensure that they remained available to communicate with their superiors, at the cost of not being available to direct the troops on the attack.[64]

The consequences of Haig's order were significant. Clearly, they limited the ability of a CO to influence the battle once it had begun, placing even greater emphasis on the battalion commander's original orders. This would have also delegated decision-making on the battlefield to more junior officers; now decisions in the minute-to-minute battle would be made by company and platoon commanders. Military historian Ian McCulloch has recorded the impacts on the battalion commander's role fairly well:

> The CO usually stayed back with the Reserve Company and HMGs [Heavy Machine Guns] or in a nearby dugout. The Battalion Commander's role in battle was to act as a sort of advanced report centre for his forward companies to report back to so information could be relayed back to the Brigadier or laterally to his flanking units. He... could manoeuvre his reserve and HMGs, request reinforcements or artillery support, liaise laterally with flanking units, or order withdrawals or realignments as required.[65]

In fairness, there was at least one good reason for the order to be given. Primarily it was because, as Bill Rawling recorded in his study of the tactics of trench warfare, "once the attack [was] launched the Battalion Commander [was] practically impotent," and unable to influence the battle any further by his presence[66] However, there was still a need for leadership on the battlefield, and senior leaders had to be seen if they were to maintain the ability to positively influence their men. This led to battalion commanders and their 2ICs often rotating between remaining in their headquarters and being up on the battlefield.[67]

Additionally, by the time of the attack on Vimy Ridge, there was one other duty that a CO was forced to perform – choosing those that would be left out of battle (LOB) to reconstitute the unit in the event of excessive casualties. Writing to his regimental namesake after the battle for Vimy Ridge, Adamson recorded both the rationale for this and the burden that it imposed:

> Since a short time before the Somme, in September [1916], G.H.Q. had decided that it was not advisable to take more than 23 officers of the Battalion into an attack. This has many advantages, particularly that of [sic] re-organising after an engagement, but its drawbacks when very valuable officers have to be left out, and they feel, after strenuous training of a Company, somewhat annoyed, but while it is hard on them, it is equally hard on the Commanding Officer having to make his selection. The regulations are that in an attack every Battalion must leave out of the trenches the Second in Command, two Company commanders and two 2nd in Command of Companies, two C.S.Ms, and 30% of specialists, and one platoon from each company...[68]

The means available to a CO to affect command were somewhat limited.[69] Radios had not entered mass usage; instead field telephones were relied upon to transmit orders and reports up and down the chain. Though effective, they had their limitations. Clearly in the offence, they were difficult to use – forcing a soldier to carry a field phone forward while laying out wire back to a battalion headquarters was an imperfect option. When a battalion was in a defensive position phones were slightly more user-friendly, however in either case, wires could be cut by shellfire and communications were never certain.

A second means was the ubiquitous "runner," a messenger that could carry verbal or written messages forward or back. This was an extremely risky role to fill on the battlefield, often exposing the chosen soldier to significant risk. Additionally, it was far from foolproof (particularly with verbal messages). Military history is replete with examples of garbled messages being given and unintended orders being carried out – often at great cost. The infamous charge of the Light Brigade during the Crimean War is a prime illustration of the issues with this method of command.

Visual means, like lamps, flags and flares, were also a tool to be used. However, they needed to be seen in order to be understood. Fog, smoke, and darkness could all make flags largely irrelevant. Similarly, the confusion of the battlefield could make flares more dangerous than not using them. Often times, when signal flares were used, the enemy could cause the signal to be misread by firing up his own coloured flares, confusing the recipient of the message.[70] As Ian McCulloch has noted, visual signalling was only good when the situation was quiet and offered little value when used in the assault.[71]

Outside of the formal written doctrine, two other means were employed to spell out how COs were to command. The first of these was the use of formal courses away from the battlefield. Adamson took one of these courses shortly after being appointed to command in October 1916. He wrote Mabel of the experience:

> There are only seven of us altogether and we are under the particular care of Brigadier General Charles... who is teaching us all what a Colonel ought to know, both by day and by night. We start at 9 a.m., the last lecturing finishing at 9.30 p.m... From 9 to 1:30 today we were laying out the ground and writing orders for the formation of a Battalion, capturing and consolidating a piece of line, which he will criticize after dinner tonight...
>
> This has been a most bitterly cold day. We have been out all day walking all over the country under General Charles, looking over the country and working out schemes for open warfare, writing orders, very interesting. The Corps Commander's idea is that next spring we will be called upon to change trench tactics to open operations and he wants the senior officers to learn it and practice it with their Battalions...[72]

Conducted near the Corps Headquarters and lasting between five and six days, the existence of these courses clearly demonstrates the interest taken by the Canadian Corps in training its battalion commanders. As Adamson's letter clearly shows, the Corps was planning well ahead. In commencing training for offensive operations a full six months before the attack on Vimy Ridge they were being proactive, anticipating the actions that they would have to take, then training their subordinate commanders for the roles that they would fill. Also noteworthy is the fact that

Julian Byng, the Commander of the Canadian Corps at the time, started Corps schools for platoon commanders to ensure that they knew their jobs equally well.[73]

The second means of instruction for COs was that given to them directly by the brigade commanders for whom they worked. In the case of Agar Adamson, this was Brigadier-General A.C. Macdonell of 7 CIB. Ian McCulloch records that "Brigadier General Macdonell was perhaps one of the most eccentric, indomitable and beloved officers to have commanded troops in the First World War."[74] Adamson was fortunate to have him as an example, writing to his wife he noted:

> I am quite pleased with our Brigadier General Macdonell, he is always on the job and seems to know his job, is most considerate... he can be depended upon to use good judgement, and not rushed into any sudden uncalled for mood without proper preparation.[75]

In addition to his example, "Batty Mac" was also quite explicit in directing his COs in the manner by which he expected them to do their jobs. While conducting research into the commander of 7 CIB, Ian McCulloch found one of Batty Mac's old notebooks where he had recorded the things that he felt important to pass on to his subordinate COs. They are enlightening to say the least. The passages were broad in scope covering tactics, command and leadership. Some of his key messages were:

> 1st eliminate self...Get ambition to do the best for [all]... Care of men, think of them all the time How they can be kept dry, fed & spared work, etc. shld [sic] be second nature...Inspection of arms shld [sic] be 2nd nature Duty of an officer ditto in Infantrymen's feet... In battle habit is everything... CO should keep in close touch with General... officers should never be glued to comd HQs but try to anticipate events by being forward. Get the habit of frequent visits at regular intervals...[76]

By the axioms he records we can see the type of commander the Brigadier was, and what he expected of his subordinates. The Brigadier preached, in essence, the adage of "mission, men, self."[77] Caring for one's men, their weapons and their health was what was important; no less so was the importance of communications higher and understanding what was happening on the battlefield. They remain as relevant today.

IN COMMAND

At the end of October 1916, Adamson was informed that he was now the CO of the Patricias, Lieutenant-Colonel Pelly having left to be an instructor at a school for COs in England.[78] After a hard summer fighting on the Somme, where Buller had been killed, the Patricias now found themselves moving into positions opposite Vimy Ridge. Agar was under no illusions as to the demands of his new job. Writing his wife, while acting CO back in August, he had penned:

> ... the isolation of a Commanding Officer is necessary but most trying to one of my disposition. One is always sir[r]ed and on the smallest points and details he has to decide and be definite and not give his reasons for doing so. One can never more than half take even senior officers into your confidence. If things go wrong it is your fault, if they go right, no matter how well thought out before, it is only considered the natural event of things. You have no idea how I yearn for a little domestic abuse from you...[79]

Adamson, affectionately nick-named "Ack-Ack" by the men of the Regiment, lost little time in putting his own stamp on the battalion.[80] Between mid-October 1916 and the attack of 9 April 1917, the Patricias served twelve tours in the line below the German positions. When they were out of the line, the unit spent its time training and preparing; after the fighting on the Somme there was much work to be done.[81] As historian Bill Rawling noted, prior to Vimy there were only 150 of the original Patricias still serving with the unit.[82] This meant that there were a significant amount of new troops, and new officers, who needed to be trained.

November 1916 passed with a series of rotations between the trenches and rest billets. The records for the period 19-21 November paint a fair picture of the period, recording "Coy parades for pay, baths, instruction in use of new gas helmets... bombing and lewis gun classes. PPCLI Comedy Coy played for the bttn [sic]... On Wednesday 22/11/16 battalion route marched to AUBIGNY."[83] Not recorded in the diary was the reason for the route march – Adamson had decided that the men needed an opportunity to buy Christmas presents for their families.[84] The next night they were back in the trenches, conducting raids, patrols and normal trench routine, getting familiar with the ground that they would be advancing across in April. December was much the same, except for one significant development.

On 9 December the Canadian Corps issued instructions on how it expected rifle platoons to be organized and employed. Rather than the four section administrative unit that platoons had been at the beginning of the war, they were to be transformed into a three section tactical organization including rifle, rifle-grenade and Lewis gun sections.[85] This was intended to optimize the hitting power of the platoon, empowering the officer in charge to act as demanded by the situation. This change was also the logical result of another observation by commanders on the battlefield. Historian David Love described the reason for the new emphasis on the platoon:

> As the war progressed, Canadian commanders recognized that due to difficulties of control in battle, the heavy losses in company commanders, and the conditions of position warfare, the battalion and company were not suitable as the basic tactical or fighting unit...[86]

On 7 January 1917, Adamson made mention of the changes in a letter to his wife writing, "The whole system of Battn. [sic] formation is being changed. It promises to have many advantages."[87]

Changes in tactics also accompanied the changes in organizational structure. These were the logical outflow of General Arthur Currie's study of French performance during the previous year. J.L. Granatstein has argued that Byng's decision to send Currie to study the French means was "the key step in changing and improving Canadian Corps tactics in the attack."[88] The intent of the new tactics was to maximize the use of artillery to "obliterate the outpost and main battle zone" held by the Germans; closely following this would be the infantry, who would seize the ground, occupy it and defeat any counter-attacks.[89] Vimy would be their first test, ultimately forming the foundation of the success that would follow in the "Last 100 Days."

German defensive tactics were concurrently undergoing their own evolution. After examining their own performance throughout 1916, German commanders shifted away from the system of "easily identifiable rigid lines" to a system of zones that were more fluid in nature. Known as elastic defence, the intent was to absorb the shock of any attack in an outpost zone, following which the Germans would use their familiarity with the ground to launch ever-larger counter-attacks through either the Rear or Main Battle Zones.[90] A good portion of the German lines had already undergone the change to elastic defence as 1917 began, however, the troops atop Vimy Ridge retained the old defensive doctrine.[91] As a result the Patricias, like every other Canadian battalion, faced the more traditional static defence, itself a formidable challenge. Historian Colonel G. W. L. Nicholson described this well:

> ... advanced fieldworks, five to seven hundred yards deep, consisted of three lines of trench with deep dugouts... interspersed by a network of concrete machine gun emplacements woven about with barbed wire, the whole system being linked by a maze of communication trenches and connecting tunnels.[92]

The Entente's strategic intent was for the French to attack north over the Oise and Aisne rivers. In the modern lexicon, they were the supported effort. The British, who would have been classed as the supporting effort, were to strike east from Arras and Bapaume. The intended combined effect of these two strikes was to rip open a gap which could then be exploited by the Entente. Field Marshall Haig decided that as part of his attacks he would also take Vimy Ridge, which dominated his northern flank.[93] As Nicholson observed, by its nature it was an attritional plan.[94] The initial orders for the Canadian Corps to start preparations were given on 2 January 1917.[95]

In the weeks that followed, the Germans conducted a general withdrawal to shorten their lines and free up forces for other needs. Although the forces at Vimy remained, much of the strategic logic for the spring campaign season was gone.[96] However, the Entente pressed forward; the attack on Vimy Ridge would occur. The Canadian plan was elegant in its simplicity, being founded on two basic ideas. As Kenneth Macksey records, they were "complete subjugation of the enemy trenches on the Ridge before the assault, and the most meticulous planning and special training of the infantry who were to take part."[97] Further, he wrote:

> The Canadian plan of attack was not in the least sophisticated. They meant to take over ground which had been neutralized by artillery fire in on straight push, all four divisions in line...[98]

While planning continued, the Patricias maintained their rotation in and out of the trenches throughout the month of January, resting and training when out of the line. Training in the new platoon tactics featured heavily.[99] This included use of a special platoon to demonstrate the new organization and its utility. A new draft of Patricias arrived to be paraded before the colours, before being inspected by Adamson and then moving into the trenches.[100] Also, on 21 January Adamson was promoted to Lieutenant-Colonel, resulting in a fair degree of celebration within the Regiment. His letter to Mabel records dinner in the mess and the general atmosphere:

> ...The Pipe Major played for a few minutes between each course... and greatly to my surprise he rose to make a speech... and then presented me with a pair of stars from the N.C.Os and men. He also asked to be allowed to play on the pipes a tune he had been composing since June, hoping for the occasion to arrive to play it to me as C.O. To me, I fear, it was very like any other pipe tune. There were some other speeches. I have had to run the gauntlet all day whenever inspecting the huts, kitchens and even on parade, and thank the men for their cheers. The spirit of approval of the appointment is, to me, most encouraging to carry on. I had no idea the great majority of men gave a damn who commanded them, but the game is not all one of cheer. Discipline has to be thought of first and this afternoon I had to have six subalterns parade for being late at the 7:30 parade this morning, and confine them to camp for two days, which seems ungrateful.[101]

In early February the Battalion was relieved of their line responsibilities to begin an intense five-week period of training prior to the expected attack in early April.[102] For the next month the focus was on training, although church, bath and pay parades also filled the time. Further, in what would be familiar to many modern soldiers, sports were used to fill time when training was not taking place.[103]

The training was thorough. The war diary records the details of attack practices, company training, bayonet fighting and grenade throwing competitions, rifle grenade practice and inspection after inspection.[104] Adamson noted on 20 February that he had been "lecturing all day to different platoons... the new attack formations are very interesting and the men are very keen to on every particular bit."[105] Concurrent to the training, the higher echelons of command continued their detailed preparations – planning the assault, stockpiling the supplies and making the arrangements that would enable success to be achieved. On 5 March, the Canadian Corps sent the final version of their plan to the Commander of the First Army for approval.[106] In general terms it detailed that the advance would be conducted behind a creeping barrage, with a series of objectives which would be facilitated by rotating the assault troops as each objective was achieved. Patricia Sergeant A.A. Bonar recorded the scenes around him:

> Everything was ready for the big show... The roads were choked with gun and ammunition limbers. The horses and mules waded through the deep muck, steaming with sweat in the sultry night. Huge ammunition lorries with throbbing motors waited by the roadside. The main roads close to the front line were lined with big naval guns and powerful howitzers... New advanced dressing stations appeared on the roadway, with motor ambulances standing quietly in front of them. Under the cover of darkness large bodies of infantry moved forward, weighted down with their heavy loads.[107]

Training for the assault also included what in modern parlance would be referred to as a "rehearsal of concept" or ROC drill. Essentially, the entire battalion got to walk through their attack on an extremely large model built behind the lines. During the rehearsal, the critical pieces of the assault were practiced including the creeping barrage, using officers on horseback with flags to simulate the pace of the advance.[108] Sergeant Bonar described the training and the thoughts behind the change in tactics:

> The ground was marked with tape to represent the enemy trenches. Various flags were placed over the ground where it was judged from aeroplane photographs were machine gun and trench mortar positions, dugout entrances, strong points and other important places of defence. Over this ground we practiced advancing under imaginary rolling barrage fire according to time table... When the barrage lifts the waves of

> men jump forward a given distance, keeping close to the barrage... The great advantage of this style of attack is apparent. The concentrated drumfire from artillery and machineguns keeps the enemy in his deep dugouts... Each platoon has a complement of 50 men made up of Lewis gun, rifle grenadier, bombing and bayonet sections, and a couple of runners and stretcher bearers. A platoon advances in two waves. Working as a unit it is most formidable in attack and able to overcome local opposition...[109]

Near the end of March, Adamson went on leave to Paris.[110] The Battalion moved into its positions opposite the German lines atop and under Vimy Ridge, in what had been named the Grange Subway.[111] Carved out of the chalk of the ridge itself, it afforded a degree of protection that would have been enviable for anyone forced to remain above ground. Ralph Hodder-Williams described the Grange tunnel system:

> 750 yards long with three exits near the jumping-off trenches and a tramline system passing close to its western entrance... the tunnel had everywhere some twenty-five feet of head cover; electric light and a water supply were installed; and there were numerous dug-outs for Battalion Headquarters, accommodation for the men, dressing stations, magazines for trench-mortars and other ammunition...[112]

Prior to the Patricias occupying their positions, work on their trenches, conduct raids and prepare for the assault, the Corps-level fireplan began. Shelling commenced on 20 March, over three weeks before the actual attack. This included more than a thousand artillery pieces, and used "over 80,000 tons of ammunition."[113] As Kenneth Macksey described, the effect of the massed artillery was devastating. German artillery was neutralized before a single infantryman moved into his attack position, communications between the forward and the depth German trenches were cut, and resupply was cut off or hindered. Even more important, tactical surprise was achieved as the gunners practiced creeping barrages on a daily basis, making it almost impossible for the Germans to know when the axe would fall.[114]

On 7 April, orders arrived from 7 CIB directing that the attack would take place, after which Adamson issued his own orders to the Battalion.[115] The next day, at

5:30 am, Batty Mac alerted his Brigade that the attack would go in on the morning of 9 April.[116] Lieutenant Lawrence Sladen, a Patricia subaltern who had been selected to participate in the assault, wrote his father, "We have been training for some time now and can't see anything but success."[117]

For the Patricias, the plan was relatively simple. In broad terms, the assault companies would move in two waves, with No. 1 Company (right) and No. 3 Company (left) tasked to seize the intermediate objectives in the area of FAMINE trench. Following the first wave would be No. 2 Company (right) and No. 4 Company (left) who would pass through the lead wave to seize the final battalion objectives approximately 300 yards further distant in the area of BRITT trench. Once both objectives were secured, the Battalion would dispatch patrols forward, and construct defensive strong points in anticipation of German counterattacks, holding until they were relieved.[118] Battalion Headquarters was to remain in the Grange subway until the first objective had been secured, after which it was to move forward into FAMINE trench.[119]

Prior to going into the assault, the men were given a hot meal and a rum ration. By 4:30 am, the Regiment was arrayed for the assault, awaiting the arrival of "zero hour."[120] The lead wave was out in the jumping-off trenches while the second wave waited their turn to go over the top in the Grange Subway. Sergeant Bonar remembered his time in the trenches opposite the Germans:

> Here we waited, our feet stuck in the gluey mud for zero time, when the order would be given to clamber over the top. Black clouds gathered and rain fell at intervals making the soil sticker [sic] than ever...Some men were leaning against the sides of the trench, others stooping or sitting in the mud on their bags of bombs. There was scarcely any conversation among the quiet crowd[121]

At 5:30 am, the assault began. Recording his thoughts almost 40 years later, Private George Hancox remembered:

> ...suddenly it was zero hour and there came the crack of what sounded like a thousand machine guns; a few seconds later came the crash of the artillery barrage. After two or three minutes, we started to go forward,

> following the rim of the crater and past the German outposts which by then were non-existent. Heavily laden we lumbered along through the maze of shell holes over the obliterated German Front Line, past more and bigger shell holes until we reached the main German Defensive line, which was the first objective...[122]

This was the first, and indeed the only, time that the Patricias would go over the top in what has become the stereotypical image of First World War combat.[123] Played over by the Regiments pipers,[124] the Battalion advanced behind the relentless barrage and took its first objectives by 7:10 am.[125] Shortly thereafter, Lieutenant-Colonel Adamson made the decision to dispatch Major McDougall forward to establish the Advanced Battalion Headquarters in FAMINE trench.[126]

Opposition up to this point in time had been relatively light, although the difficulties on the Division's northern flank where Hill 145 had not been taken would cause increasing casualties as the day wore on.[127] By 9:45 am, the Patricias were firm on both of their objectives and patrols were dispatched forward, while runners were sent back to request more small arms ammunition and hand grenades.[128] Telephone communications were then established between Advanced and Main Battalion Headquarters at around 11:30 am.[129] Less than half an hour later, the CO left the Grange Subway and moved forward to take up a position in FAMINE trench; sending Major McDougall again forward to the area of BRIDLE trench to establish a new advanced headquarters.[130] Throughout the rest of the day, fighting continued within the Patricia area of responsibility as adjustments were made to cover flanks that had opened up as a result of stronger resistance in the vicinity of Hill 145. Although things had gone remarkably well, the Regiment suffered just over 200 casualties.[131] Among the dead was Lieutenant Sladen, killed during the initial assault.[132]

After months of preparations, the Patricias achieved success in less than seven hours. In return for their efforts, the Patricias were treated to a view of the Douai Plain. Private Hancox was awestruck when he saw the other side of the ridge:

> After a short interval on the final objective, which was just over the crest of the hill, and there I saw the vast expanse of the country beyond. Redbrick mining villages among a score of slagheaps and pitheads.

> Liven, Avion, Mericourt and Lens itself and other villages farther back. In the foreground were Vimy and Petit Vimy. All appeared to be untouched... After months in the wilderness it was truly a glimpse of the promised land.[133]

Lieutenant-Colonel Adamson's actions on 9 April are not well recorded. His letters to Mabel from 10 to 12 April 1917 record some significant impressions and some specifics of what the Regiment did, but precious little about what he did in particular. Similarly, the war diary does not record much beyond some of the decisions that he made, such as when to move his headquarters. There is one account of him and his batman going souvenir hunting on the Ridge, finding some German tonic water that Adamson thought would go well with Scotch, but there is no definitive account of what he did throughout the entirety of the action.[134] Instead we are forced to make an educated guess, based on the doctrine that was in use, and the tasks that the Patricias were given, in order to discern what types of things he would have been concerned about as the attack took place.

As a CO, Adamson would have been thinking about four broad areas: what was going on in his sector and around him, the Regiment's ability to keep fighting, what was going to occur next, and his men. Reports from the battlefield would have been coming in by runner or telephone into the Headquarters. These would have detailed what the companies were doing and seeing, problems they were having, and requests for assistance. Sorting through these and trying to get an accurate picture of what was occurring would have been a principle concern. Similarly, Adamson would have been trying to understand what was going on both to his left and right, where the Royals and the Blackwatch each fought their own battles. Close coordination would have been required to make sure that opportunities were not given to the Germans, such as exposing flanks or leaving gaps in defensive fires. Additionally, this information would have been required at the Headquarters of 7 CIB.

Keeping the Battalion in the fight would also have taken a fair amount of his efforts. Things like making sure casualties were being evacuated, that mopping-up parties were moving forward, and that prisoners were being sent to the rear would all have needed attention. Additionally, he would have wanted to confirm that water and ammunition were moving forward to the companies. Knowing that strong points were to be constructed would also have meant that Adamson would have needed to ensure that the companies had sufficient defensive stores (like wire, tools and sandbags) to do the job.

The careful movement of Battalion Headquarters further forward would also have been a significant issue. Moving a headquarters makes it vulnerable, and there would have had to have been confirmation between Adamson and McDougall as to who was moving when, and who had control over the Battalion as the fight went on. Lastly, he would have been particularly interested in the status of the machine guns and trench mortars, which were being used by the companies. These two weapon systems were the only heavy weapons that the Battalion had at its immediate beck and call; it was vital that they be ready for action if a counter-attack did occur.

Understanding that casualties were being taken from the German positions around Hill 145, it is logical to deduce that Adamson would have been extremely interested in what was taking place there and what actions were being planned to seize that position. It is also fair to assume that he would have been interested in knowing when the Canadian artillery was going to be moving forward. Likely, he would have also wanted to know how long his men were expected to hold their positions. Additionally, he would have wanted to know that he had accounted for them all: who had been wounded or killed and who was missing.

After having spent just over a day on the Ridge, Adamson rotated his lead and depth companies in a bid to give the men on the front lines some respite.[135] Less than an hour later, orders were received from 7 CIB; the Patricias were to be replaced in the line and would be moving into the Brigade Support Area. By 8:30 pm on 11 April 1917, this was underway.[136] Operations on Vimy Ridge, for Agar Adamson and the Patricias, were done. In the days following the attack, he wrote his wife Mabel. Included in one of the letters was "a primrose and a violet picked in the woods in front of the ridge."[137]

Vimy was, in many ways, the high point of his military service. After the battles of Easter Sunday 1917, Adamson continued in command and again led the Battalion in battle in the meat grinder of Passchendaele. By the time the Regiment came out of that battle, there was only one other "original" officer left – Charlie Stewart, who would later replace Adamson as CO.[138] This in itself a perpetuation of both the formal and informal training that Adamson himself had gone through.

Adamson continued in command until early January 1918 when, as Ralph Hodder-Williams records, he "ended his long career with the Regiment in the field, having been pronounced by a medical board unfit for further front-line service."[139] For Adamson the war was essentially over, although he went on to become a "District Commander" working for the Canadian Corps.[140]

AFTER THE FIGHTING

Six years after the First World War ended, some of the hard-earned battlefield lessons were encapsulated in a rewrite of the *Field Service Regulations* (FSR).[141] If one looks upon their contents as the best practices witnessed during the war, it is insightful to review what they said. As regards commanders and their role, the FSR were quite specific stating "the chief role of a commander is to make decisions... Commanders of all ranks must be imbued with the doctrine that inaction and fear of responsibility are graver faults than errors in the choice of plan."[142] Going further, the doctrine asserted:

> [H]eadquarters... should be so placed as to enable the commander to keep in close and constant touch with the fighting troops, with his reserves, and with the other arms supporting him, and if necessary, to intervene personally in the conduct of the battle...
>
> The headquarters of an attacking unit must be established well forward from the beginning of the battle, and must be moved forward by bounds in order to keep in touch with the troops as they advance...
>
> In all cases the commander should show himself to his troops whenever opportunities occur...[143]

By this time however, Agar Adamson had left the military. His time in command had left scars far beyond the physical.[144] Sadly, his marriage did not survive the war and he spent much of his postwar life in Ottawa and England, while Mabel remained in Port Credit, Ontario.[145] In October 1929, after taking up an interest in flying, he attempted an over-flight of Ireland with a British aviator. Although he survived the crash into the Irish Sea, he was left exposed to the ravages of the elements for over two hours. He died on the 21st of November 1929, attended by his wife, and son Anthony who had travelled to his bedside.[146]

Vimy Ridge is now still and quiet, with the exception of the tourists who flock to see the Canadian memorial at the site of Hill 145. It is still possible to see the opposing trench lines in the Patricia sector and imagine the men going over the top, sleet raining down as they lean into the barrage and follow it forward. The old battalion headquarters and Adamson's quarters are still there, dug into the chalk of the ridge; it is possible to go and see the room to which the runners would have breathlessly returned so that reports could have been sent up to Batty Mac at 7 CIB. Off to the side there is a small squarish room cut out of the chalk. It contains a small chicken wire covered bed frame, a wooden shelf and a writing table – complete with an old tin cup and the remnants of a lamp. The soft amber coloured light belies the importance of the actions that were planned and controlled from these two rooms. One can almost imagine Agar Adamson sitting there still.

---

1 Ralph Hodder-Williams, *Princess Patricia's Canadian Light Infantry 1914-1919, Second Edition.* (Toronto: Carswell Printing, 1968), p.221.

2 In the case of the PPCLI, the terms Unit, Regiment and Battalion should be taken to all mean the same thing. From their formation in 1914 until the commencement of the Korean War, the PPCLI possessed only a single battalion. Further details on their organizational structure will be covered later in this chapter. By convention, the unit is also known as the "Patricias."

3 As recorded by Pierre Berton, there were 29 Battalions who went "over the top" on 9 April 1917. Additionally there were a further 19 Battalions in a reserve role, thus 48. Pierre Berton, *Vimy* (Toronto: McClelland and Stewart, 1986), pp.310-312.

4 The units (from right to left) were the Royal Canadian Regiment, the PPCLI, and the 42nd Battalion (the Black Watch). The remaining Brigade unit was the 49th Battalion from Edmonton, which was split between the assaulting units to act as mopping up parties and as the Brigade reserve.

5 The term Commanding Officer is the modern title for the senior commander in an infantry battalion. During the First World War, the title for a battalion commander was actually "Officer Commanding." To limit confusion, I will use the term CO throughout when referring to a unit commander.

6 In his popular history of the battle, for example, author Pierre Berton pays relatively little attention to the Battalion Commanders that led their units in battle. He specifically notes that the book was not intended "to be definitive from a military or tactical point of view. I have not thought it necessary to mention every battalion that took part in the action or every senior officer... I have tried to look at the Vimy experience from the point of view of the man in the mud as well as the senior planners." Berton, *Vimy...*, p.313.

7 Department of National Defence, *Leadership in the Canadian Forces: Conceptual Foundations* (Ottawa: DND Canada, 2005), p.7.

8 Ibid., p.8.

9 Similarly, though outside the scope of this chapter, it needs to be acknowledged that one can lead without the formal appointment of being a commander.

10 See Department of National Defence, *Leadership in the Canadian Forces: Conceptual Foundations* (Ottawa: DND Canada, 2005).

11 Sandra Gwyn, "Dictionary of Canadian Biography Online: Agar Stewart Allan Masterton Adamson," <http://www.biographi.ca/EN/ShowBio.asp?BioId=42135&query=Agar%2 AND %20Adamson>; Internet; accessed 10 April 2008.

12 Ibid.

13 Gwyn, "Dictionary of Canadian Biography Online: Agar Stewart Allan Masterton Adamson." and Sandra Gwyn, *Tapestry of War: A Private View of Canadians in the Great War* (Toronto: Harper Collins, 1992), p.75.

14 Gwyn, "Dictionary of Canadian Biography Online: Agar Stewart Allan Masterton Adamson."

15 Gwyn, *Tapestry of War...*, pp.76-77. The pairing was fortunate from a historical standpoint as it lead to a detailed correspondence which would last throughout their lives, to later be deposited in the National Archives by their son. This wealth of material forms a primary source for this and numerous other projects by a wide variety of authors across Canada.

16 Ibid., pp.78-79.

17 Gwyn, "Dictionary of Canadian Biography Online: Agar Stewart Allan Masterton Adamson."

18 Craig Leslie Mantle, *Learning the Hard Way: The Leadership Experiences of Lieutenant Agar Adamson During the South African War, 1899-1902* (Kingston: Canadian Defence Academy Press, 2008), p.6. Mantle's work is an impressive piece of scholarship that clearly chronicles the earliest active military experiences enjoyed by Agar. It is highly recommended to any wishing to get a more complete understanding of this period of Adamson's life.

19 Mantle, *Learning the Hard Way...*, pp.28-36.

20 Ibid., p.64.

21 Ibid., pp.70-71.

22 Gwyn, "Dictionary of Canadian Biography Online: Agar Stewart Allan Masterton Adamson."

23 Gwyn, *Tapestry of War...*, p.71.

24 Patrick H. Brennan, "Good Men For a Hard Job: Infantry Battalion Commanders in the Canadian Expeditionary Force," *Canadian Army Journal* 9,1 (Spring 2006), pp.10-11.

25 Mantle, *Learning the Hard Way...*, p.68.

26 Ibid.

27 Ian McCulloch, ""Batty Mac": Portrait of a Brigade Commander of the Great War. 1915-1917," *Canadian Military History* 7, 4 (Autumn 1998): 12.

28 Adamson's choice of regiments may have been because he expected that the Canadian Militia would not have enlisted him as one of his eyes was blind. David J. Bercuson, *The Patricias: The Proud History of a Fighting Regiment* (Toronto: Stoddart, 2001), p.25.

29 There is no record of Adamson and Gault meeting during the Boer War. Indeed, Sandra Gwyn has asserted that the two men were not acquainted with one another when Gault was forming the PPCLI. Gwyn, *Tapestry of War...*, p.83.

30 Hodder-Williams, *Princess Patricia's Canadian Light Infantry 1914-1919...*, p.9.

31 Ibid., p.10.

32 Ibid., p.17.

33 Bercuson, *The Patricias...*, p.30.

34 Ian McCulloch, ""The Fighting Seventh": The Evolution and Devolution of Tactical Command and Control in a Canadian Infantry Brigade of the Great War." (Master's thesis, Royal Military College, 1997), p.34.

35 Information for this organizational model is a compilation of Bercuson, *The Patricias...*, p.30, David W. Love, *"A Call to Arms": The Organization and Administration of Canada's Military in World War One* (Winnipeg: Bunker to Bunker Books, 1999), 30-31 and McCulloch, ""The Fighting Seventh"..., pp.70-71.

36 Hodder-Williams, *Princess Patricia's Canadian Light Infantry 1914-1919...*, pp.18-19.

37 Gwyn, *Tapestry of War...*, pp.118-119.

38 Farquhar's subsequent guidance to Adamson may be seen at Robert F. Zubkowski, *As Long As Faith and Freedom Last: Stories From the Princess Patricia's Canadian Light Infantry From June 1914 to September 1919* (Calgary: Bunker to Bunker Books, 2007), pp.52-53.

39 N.M. Christie (Editor), *Letters of Agar Adamson* (Nepean, Ontario: CEF Books, 1997), pp.16-17 (letter dated 9 Feb 1915).

40 Zubkowski, *As Long As Faith...*, p.53.

41 Christie, *Letters of Agar Adamson...*, p.39 (letter dated 10 March 1915).

42 Ibid., p.45 (letter dated 21 March 1915).

43 Hodder-Williams, *Princess Patricia's Canadian Light Infantry 1914-1919...*, p.44.

44 Christie, *Letters of Agar Adamson...*, p.59 (letter dated 11 April 1915).

45 Ibid., p.60 (letter dated 12 April 1915).

46 Ibid., p.73 (letter dated 5 May 1915) and Hodder-Williams, *Princess Patricia's Canadian Light Infantry 1914-1919...*, pp.56-57.

47 Hodder-Williams, *Princess Patricia's Canadian Light Infantry 1914-1919...*, p.47.

48 Ibid, pp.60-65.

49 Ibid., p.68. The PPCLI War Diary records that Adamson's partner in distributing ammunition, Sergeant-Major Fraser, was killed while they worked together. Library and Archives Canada (LAC), "War diaries – Princess Patricia's Canadian Light Infantry – 1917," RG 9, D, III, Volume 4912, 8 May 1915 10:30 AM. Hereafter referred to as War Diary. National Archives Canada has placed a substantial number of unit war diaries from the Canadian Expeditionary Force on-line. They may be accessed at <http://www.collectionscanada.gc.ca/archivianet/020152_e.html>.

50 Hodder-Williams, *Princess Patricia's Canadian Light Infantry 1914-1919...*, p.70.

51 Ibid., p.71.

52 Hodder-Williams, *Princess Patricia's Canadian Light Infantry 1914-1919...*, p.76.

53 See Gwyn, *Tapestry of War...*, pp.156-158.

54 Hodder-Williams, *Princess Patricia's Canadian Light Infantry 1914-1919...*, pp.87-89.

55 Jeffery Williams, *First in the Field: Gault of the Patricias* (St Catharines, Ontario: Vanwell, 1995), p.118. William's comments on Adamson's confidence in his abilities is likely explained by having served approximately a year in the trenches and seeing numerous friends wounded or killed, including his most recent CO.

56 War Diary, 3 August 1916. There is some debate as to whether or not Adamson wished to command, or wished to fill another role in the Regiment. Historian David Bercuson has made much of the fact that in the Regimental History, Ralph Hodder-Williams wrote that Lieutenant-Colonel Pelly "was allowed to rejoin his old Regiment at the request of Major Agar Adamson" (Bercuson, *The Patricias...*, pp.86-87 and Hodder-Williams, *Princess Patricia's Canadian Light Infantry 1914-1919...*, p.145). Bercuson asserts that "Adamson was reluctant to continue [in command]" (Bercuson, *The Patricias...*, p.86). Adamson's letters offer some insight into his mindset. His letter to Mabel, of 27 July 1916, explains that the Regiment was very short of senior officers, hence Adamson's outreach to Pelly. Pelly was senior to Adamson, having been appointed a Major when the unit was formed. In some ways it was only natural for Adamson to expect to turn the Regiment over to him. Agar closed the letter to Mabel with a telling statement "I am not thinking of myself, but only acting for what I think will prove best for the Regiment and for our side"(Christie, *Letters of Agar Adamson...*, p. 201 (letter dated 27 July 1916)). A belief in a Regimental order of seniority is also evident in a letter to Adamson written by Talbot Papineau in May 1917 wherein he states "I have heard that Charlie [Stewart, the Regimental Second in Command] has been declared unfit for further service. This may or may not be so. I hope not, but if this is so, I presume that I am now due as second in command" (Letter dated 10 May 1917, LAC) "Agar Adamson Fonds," MG 30, Series E149, May 1917). The fact that Agar does not appear to have attempted to get out of front line service, and would ultimately serve over 30 months in the trenches, tends to disprove Bercuson's idea that he was not up to the physical

exertions of command. As to his mindset or his motivation for asking that Pelly take command, it is all conjecture.

57 Patrick Brennan and Thomas Leppard, "How the Lessons Were Learned: Senior Commanders and the Moulding of the Canadian Corps After the Somme," in *Canadian Military History Since the 17th Century: Proceeding of the Canadian Military History Conference*, Ottawa, 5-9 May 2000, edited by Yves Tremblay, pp.135-143 (Ottawa: Directorate of History and Heritage, 2000), p.136.

58 The other two publications were General Staff, War Office (GSWO), *Field Service Pocket Book. 1914* (London: His Majesty's Stationery Office, 1914) and GSWO, *Field Service Regulations, Part I Operations 1909 (Reprinted with Amendments 1912)* (London: His Majesty's Stationery Office, 1912).

59 GSWO, *Infantry Training (4-Company Organization) 1914* (London: His Majesty's Stationery Office, 1914. Reprint Uckfield, East Sussex: The Naval & Military Press, 2006). All page numbers will be as taken from the reprint edition.

60 Ibid., p. 121.

61 GSWO, Infantry Training..., p.141.

62 Ibid., p.140.

63 Ibid., p.121.

64 M.A. Ramsey, *Command and Cohesion: The Citizen Soldier and Minor Tactics in the British Army, 1870-1918* (Westport, Connecticut: Praeger, 2002), p.178.

65 McCulloch, ""The Fighting Seventh"..., p.22.

66 The words were actually those of a CO of the Royal 22nd Regiment (the Vandoos) in 1916. Bill Rawling, *Surviving Trench Warfare: Technology and the Canadian Corps, 1914-1918* (Toronto: University of Toronto Press, 1992), p.81.

67 McCulloch, ""The Fighting Seventh"..., p.36.

68 Letter to Princess Patricia, dated 17 April 1917. LAC, "Agar Adamson Fonds," MG 30, Series E149, April 1917. The Somme battles that Adamson refers to were the costly Canadian efforts of the fall, and not the attacks of that summer.

69 Information for this section is from Rawling, *Surviving Trench Warfare*..., pp.85-86.

70 As a new platoon commander I learned this from personal experience. Before going out on patrol I gave orders on actions to be taken in the event of different coloured flares being fired. It was at this point that I was informed that one of my section commanders was colour blind and would have to be paired with someone who could tell the difference between green and red.

71 McCulloch, ""The Fighting Seventh"..., p.36. The same could be said of carrier pigeons, whose utility largely depended on them not being confused by the noise of the battlefield.

72 Christie, *Letters of Agar Adamson*..., pp.234-235 (letters dated 16 and 17 November 1916).

73 Jeffery Williams, *Byng of Vimy: General and Governor General* (Toronto: University of Toronto Press, 1992), p.147.

74 McCulloch, ""The Fighting Seventh"..., p.56.

75 Christie, *Letters of Agar Adamson...*, p.131 (letter dated 20 January 1916).

76 As cited in McCulloch, ""The Fighting Seventh"..., pp.58-59. This is also referred to in Chapter 3, p. 56.

77 First heard by the author, as an officer cadet, from an old Patricia Warrant Officer.

78 Hodder-Williams, *Princess Patricia's Canadian Light Infantry 1914-1919...*, p.194.

79 Christie, *Letters of Agar Adamson...*, p.203 (letter dated 2 August 1916).

80 Gwyn, *Tapestry of War...*, p.424. Some sources also cite "Ackety-Ack." See Zubkowski, *As Long As Faith...*, p.272. P.H. Ferguson was one of Adamson's runners and does not appear to have cared for some of the Colonel's eccentricities. He seems to have been one of the few soldiers who did not like the man, unlike several other soldiers who record Agar's soft touch when disciplining them.

81 Hodder-Williams, *Princess Patricia's Canadian Light Infantry 1914-1919...*, p.191.

82 Rawling, *Surviving Trench Warfare...*, p.97.

83 War Diary, 18-22 November 1916.

84 Christie, *Letters of Agar Adamson...*, p.236 (letter dated 22 November 1916). The letter records some of the men's actions, who chose to get drunk instead. It also records some of the actions that Adamson took, anticipating that soldiers being given a break from routine might be apt to overindulge.

85 Brereton Greenhous and Stephen J. Harris, *Canada and the Battle of Vimy Ridge: 9-12 April, 1917* (Ottawa: Minister of Supply and Services, 1992), p.58.

86 Love, "*A Call to Arms*"..., p.32.

87 Christie, *Letters of Agar Adamson...*, p.252 (letter dated 7 January 1917).

88 J.L. Granatstein, *Canada's Army: Waging War and Keeping the Peace* (Toronto: University of Toronto Press, 2002), p.110.

89 Shane B. Schreiber, *Shock Army of the British Empire: The Canadian Corps in the Last 100 Days of the Great War* (Westport, Connecticut: Praeger, 1997), p.11.

90 John A. English, *On Infantry* (New York: Praeger, 1981), pp.15-16. This book is highly recommended to anyone who seeks to understand the evolution of infantry tactics over the course of the 20th Century.

91 Brereton Greenhous and Stephen Harris have argued that there were three main reasons that the Germans atop Vimy Ridge did not adopt the tactics of elastic defence. First, their defences had already been proven effective in the face of British and French attacks. Second, a significant amount of work had already been done. Adopting the new techniques would have meant throwing that work away on a largely unproven concept. Third, they felt that the ground did not support a doctrine based on counter-attacks as the northeastern side of the ridge dropped steeply away onto the Douai Plain. Greenhous and Harris, *Canada and the Battle of Vimy Ridge...*, pp.68-70. One cannot help but note that the Germans did not see counte-attacks as practical,

whereas the Canadians devoted great energy to being able to face heavy counter attacks once they had wrested Vimy Ridge away from the Germans.

92 Colonel G.W.L. Nicholson, *Canadian Expeditionary Force 1914-1919: Official History of the Canadian Army in the First World War* (Ottawa: Queens Printer, 1962), p.246.

93 Kenneth Macksey, *The Shadow of Vimy Ridge* (Toronto: Ryerson, 1965), pp.52-54.

94 Nicholson, *Canadian Expeditionary Force 1914-1919...*, p.244.

95 Macksey, *The Shadow of Vimy Ridge...*, p.57.

96 Ibid., pp.53-54.

97 Macksey, *The Shadow of Vimy Ridge...*, p.65.

98 Ibid., p.67.

99 War Diary, 19-23 January 1917.

100 Ibid., 23 January 1917.

101 Christie, *Letters of Agar Adamson...*, p.256 (letter dated 21 January 1917). Adamson's actions with regard to his subalterns warrants note not only because it demonstrates his belief in discipline within the unit, but also because when placed in the context of being out of the line for a six day period, ordering one's men to remain on camp was a fairly strong punishment.

102 War Diary, 7 February 1917.

103 In a series of interesting War Diary entries, it becomes apparent how important football (soccer) was to the unit. The entry for 13 February records that they played in the semi-finals for the Brigade against the 49th Battalion. No record is made of the victor, although evidence seems to point to a Patricia loss. Later on 1 March, the war diary records an officer's match against the RCR. Here it duly records that the officers won the contest against the Royals. Hodder-Williams in his history of the Regiment notes specifically that the 42nd Battalion "excelled" in this particular skill set. War Diary, 13 February and 1 March 1917 and Hodder-Williams, *Princess Patricia's Canadian Light Infantry 1914-1919...*, p.207.

104 War Diary, February and March 1917.

105 Christie, *Letters of Agar Adamson...*, p.265 (letter dated 20 February 1917).

106 Nicholson, *Canadian Expeditionary Force 1914-1919...*, p.244.

107 Zubkowski, *As Long As Faith...*, pp.257-258.

108 Macksey, *The Shadow of Vimy Ridge...*, p.66.

109 Zubkowski, *As Long As Faith...*, pp.251-252.

110 War Diary, 21 March 1917.

111 War Diary, 22 March 1917.

112 Hodder-Williams, *Princess Patricia's Canadian Light Infantry 1914-1919...*, pp.211-212.

113 Macksey, *The Shadow of Vimy Ridge...*, p.65.

114 Ibid., p.70 and Rawling, *Surviving Trench Warfare...*, p.112. The number of artillery pieces allowed for one gun for every 20 metres of frontage for the assault.

115 A copy of Adamson's orders were kept with the Unit War Diary for April 1917(Appendix B). Additionally, a "narrative of operations" was written, detailing the minute to minute events covering the period 8-11 April 1917 (Appendix C).

116 War Diary, 8 April 1917. The military at this time had not adopted the 24 hour system of recording time. They used the familiar AM and PM instead.

117 Letter from Lawrence Sladen, dated 7 April 1917, LAC, "Agar Adamson Fonds," MG 30, Series E149, April 1917. Lawrence Sladen's father, Arthur, was a friend of Agar Adamson's who worked in the Privy Council in Ottawa. Arthur had facilitated Agar's introduction to Hamilton Gault and played a part in Adamson joining the PPCLI.

118 War Diary, April 1917, Appendix B, 10. The task of holding until they were relieved is implied by the tone of the order, their assigned tasks of building strong points and capturing a section of the ridge.

119 Ibid.

120 Hodder-Williams, *Princess Patricia's Canadian Light Infantry 1914-1919...*, p.217.

121 Zubkowski, *As Long As Faith...*, p.263.

122 G.T. Hancox, *Historical Notes*, as cited in Newman, *With the Patricia's Capturing the Ridge...*, p.50.

123 Jeffery Williams, *Princess Patricia's Canadian Light Infantry: 1914-1918 Seventy Years of Service* (London: Leo Cooper, 1985), p.22.

124 Hodder-Williams, *Princess Patricia's Canadian Light Infantry 1914-1919...*, p.218.

125 War Diary, April 1917, Appendix C, 13.

126 Ibid.

127 Greenhous and Harris, *Canada and the Battle of Vimy Ridge...*, p.104.

128 War Diary, April 1917, Appendix C, 13.

129 Ibid.

130 Newman, *With the Patricia's Capturing the Ridge...*, p.58.

131 War Diary, April 1917, Appendix C, 14. Approximately 25% of these occurred during the actual assault. The vast majority occurred in the 36 hours which followed, while the Patricias were holding their position and exposed to fire from Hill 145 which was not taken in the initial assault.

132 Letter to Princess Patricia dated 17 April 1917, LAC, "Agar Adamson Fonds," MG 30, Series E149, April 1917.

133 Zubkowski, *As Long As Faith...*, p.271.

134 The story itself is entertaining and offers a different perspective on the man. Zubkowski, *As Long As Faith...*, p.272.

135 War Diary, April 1917, Appendix C, 15.

136 Ibid, p.16.

137 Gwyn, *Tapestry of War...*, p.351.

138 Hodder-Williams, *Princess Patricia's Canadian Light Infantry 1914-1919...*, pp.283-284. Stewart had started the war as a lieutenant with the Regiment.

139 Hodder-Williams, *Princess Patricia's Canadian Light Infantry 1914-1919...*, p.283.

140 Christie, *Letters of Agar Adamson...*, p.333 (letter dated 30 March 1918).

141 General Staff, War Office, *Field Service Regulations, Volume II* (Operations) 1924 (London: His Majesty's Stationery Office, 1924).

142 Ibid., pp.9-10.

143 *FSR...*, p.42.

144 Gwyn, "Dictionary of Canadian Biography Online: Agar Stewart Allan Masterson Adamson."

145 Gwyn, *Tapestry of War...*, pp.430-432.

146 Gwyn, "Dictionary of Canadian Biography Online: Agar Stewart Allan Masterson Adamson."

CONTRIBUTORS

Patrick Brennan has been a faculty member in the History Department at the University of Calgary since 1989. He earned his PhD from York University (Toronto) and his MA from the University of Regina. He is also a Fellow of the Centre for Military and Strategic Studies at the University of Calgary. His area of research interest is the Canadian Expeditionary Force in the First World War. Among other projects, he is currently working on a monograph entitled *Byng's and Currie's Commanders: A Study of Leadership in the Great War.*

David Campbell completed his PhD in history at the University of Calgary where he specialized in military history. His major area of research is the social and operational history of the Canadian Expeditionary Force during the First World War. He currently resides and teaches in Halifax, Nova Scotia.

Paul Dickson is a strategic analyst with the Centre for Operational Research and Analysis, currently posted to the Directorate of Air Strategic Plans. His most recent work is *A Thoroughly Canadian General: A Biography of General H.D.G. Crerar,* published by the University of Toronto Press.

Andrew B. Godefroy is a strategic analyst and historian with the Canadian Army's Directorate of Land Concepts and Designs, as well as the editor in chief of the *Canadian Army Journal.* He has served with the Canadian Army at all levels of command, and is a graduate of the Canadian Forces School of Military Engineering as well as the Canadian Land Forces Command and Staff College. He holds both an MA and PhD in War Studies from the Royal Military College, Kingston, as well as a visiting research fellowship in the Leverhulme Programme on the Changing Character of War at Oxford University. He has authored over a dozen studies on the Canadian Expeditionary Force, and is currently completing a new biography of Lieutenant-General Sir Arthur Currie.

Ian McCulloch is a native of Halifax, Nova Scotia. Educated in Scotland and Switzerland, he holds a degree in Journalism from Carleton University and a Master's in War Studies from the Royal Military College of Canada. He joined the

Canadian Army in 1977 and has served in a variety of regimental and staff appointments in Canada, USA and Germany. Promoted to lieutenant-colonel in 1993, he assumed command of the Black Watch (RHR) of Canada (1993-96) and from 1996 to 2000 was Deputy Director of History & Heritage for the CF in Ottawa. Since October 2007, he has been a member of the Directing Staff at the Canadian Forces College, Toronto. Lieutenant-Colonel McCulloch is an avid military historian specializing in the Seven Years War in North America and has published numerous articles and monographs on that subject as well as several books. His most recent work is the two volume series, *Sons of the Mountains: A History of the Highland Regiments in North America, 1756-1767.*

David O'Keefe is a Gemini-nominated historian with a dozen television documentaries to his credit including *Black Watch: Massacre at Verrieres Ridge.* Research on all aspects of the Black Watch started 15 years ago when O'Keefe was an officer in that regiment and an undergraduate in history at McGill and Concordia Universities. After leaving the Army to continue his academic career, O'Keefe was brought onboard as the signals intelligence specialist by the Directorate of History and Heritage during the writing of the official history of the Royal Canadian Navy in the Second World War. In 2003, he returned to the Black Watch in the position of Regimental Historian and has held teaching positions at John Abbott College in Montreal as well as the University of Ottawa.

Tod Strickland is a serving infantry officer with the Canadian Army. A member of the Princess Patricia's Canadian Light Infantry since 1989, he has served in Bosnia, Croatia, and Afghanistan along with participation in several domestic operations. He is a graduate of the Canadian Forces Joint Command and Staff Program, and has earned a Masters of Arts (History) from American Military University, as well as a Masters in Defence Studies and a Bachelor of Military Arts and Sciences from the Royal Military College.

Timothy C. Winegard is currently completing his Doctor of Philosophy in History at the University of Oxford under Professor Hew Strachan. His thesis is a comparative study on the use and treatment of the Indigenous populations from Canada, Australia, New Zealand and South Africa during the First World War. His M.A. thesis at the Royal Military College of Canada focused on the Oka Crisis and the Canadian Forces. Timothy is an Officer with the 1st Hussars Reserve Armoured Regiment and is currently attached to the 7 Rifles Battalion of the British Territorial Army.

INDEX

ENTENTE UNITS

IMPERIAL GERMAN ARMY